Technology and Communication Behavior

Technology and Communication Behavior

Frederick Williams

Center for Research on Communication Technology and Society
The University of Texas at Austin

With contributions from Milton Chen, Herbert S. Dordick, William H. Dutton, Martin Elton, Liz Greenberger, Patricia M. Greenfield, Andrew P. Hardy, Heather E. Hudson, Suzanne Iacono, Rob Kling, William Paisley, Pamela Pease, Amy F. Phillips, Ronald E. Rice, Everett M. Rogers, Paul A. Strassmann, Charles Steinfield, and Victoria Williams

Wadsworth Publishing Company
Belmont, California
A Division of Wadsworth, Inc.

Mass Communication Editor: Kristine Clerkin
Editorial Assistant: Naomi Brown
Production Editor: Harold Humphrey
Print Buyer: Karen Hunt
Designer: Lisa S. Mirski
Copy Editor: Yvonne Howell
Compositor: Kachina Typesetting
Cover: Mark McGeoch

Printed in the United States of America

1 2 3 4 5 6 7 8 9 10 91 90 89 88 87

Library of Congress Cataloging in Publication Data
Williams, Frederick, 1933–
  Technology and communication behavior.

  Includes bibliographies and index.
  1. Telecommunication—Social aspects. I. Chen, Milton, II. Title.
HE7631.W55   1987   303.4'833   86-15962
ISBN 0-534-07398-0

Preface

A fundamental premise of *Technology and Communication Behavior* is that in this period approaching the 21st century, we are experiencing a "settling in" of the communication technology revolution. Although the technological innovations may continue to impress us, it is ultimately the impact upon people that is the real and continuing change that we sense and to which we are challenged to adjust. In the last decade the glitter and often unbridled promises of such technologies as personal computers, the communications satellite, cable television, video disk and video cassette, high-resolution television, office automation devices, videotext and teletext, and electronic mail have given way to the realities of the marketplace. More specifically, the tests have been ones of human and technology integration as they have occurred in the modern office, the home, or our service institutions of education, health, or defense.

To promote an understanding of these human and communication technology interactions, to be able to evaluate them, and ultimately to gain the maximum benefits for society is the aim of Technology and Communication Behavior.

The present volume has been written, first, as a textbook for classes or seminars where the focus is upon the social and behavioral aspects of communication technologies. As such, its contents have been organized and various features added to enhance the pedagogical value of the presentation. This is in contrast to the already existing (some excellent) edited volumes of readings on communications technologies. Accordingly, there is an attempt in *Technology and Communication Behavior* to provide an orientation in technological applications and to promote an understanding of them, not just to review the literature. Part I is devoted to introducing this orientation.

A practical matter in the study of modern communications is some fundamental understanding of the underlying components of new media. Part II examines telecommunications and computing. The coverage is generally nontechnical yet gives insights into the basic components and their functions.

The details of human and technology interaction, however, are best found in examples of contemporary research, which form the topics of the seven chapters of Part III. This literature is not so much offered as a survey of *all* research on the matter, but more as an attempt to present a comprehensive view of the wide range of applications environments—from the individual to the group, organization, and national environments. Toward this end, experts in the field have contributed brief reports on research topics in their areas of specialization. More than simply reviewing pertinent literature, the goal is to interpret it and offer generalizations that transcend particular studies and point to our future.

Among the trends in communication study observed early in this volume is that change itself is an increasing topic of research and theoretical generalization. This concept is elaborated in the two chapters of Part IV on gratifications theory and new research foci.

Communication technologies and behavior are perhaps the most rapidly evolving topics in the communication field. I have concentrated on examples and generalizations that transcend particular instances of application. I believe that I have made useful choices for the reader. Where I have not, I hope that you will assist by sharing your criticisms. In that manner we can work together to advance the literature of our field.

I am grateful for the comments and suggestions of the several reviewers who consented to read an earlier draft of the manuscript. These reviewers included Kent Creswell, Michigan State University; James Danowski, University of Illinois, Chicago; and Mitchell L. Moss, New York University.

Finally, may I add a special acknowledgement to those scholars, listed on the title page, who contributed the special "Reports" that appear in Part III, and a further special thanks to friends and colleagues Ronald Rice, Martin Elton, Charles Steinfield, William Dutton, and Sharon Strover, who contributed ideas for discussion and research exercises.

Frederick Williams
Center for Research on Communication Technology and Society
The University of Texas at Austin

Contents

PART II · TECHNOLOGICAL ADVANCES / *43*

1

THE INFORMATION REVOLUTION

In the first three chapters of this volume we examine communication technologies from the broadest views. Chapter 1 introduces the concept of change. Chapter 2 proposes an analytic approach to understanding communication processes. Chapter 3 provides an overview of selected communication technologies and their functions in society.

Communication, Technology, and Social Change

The technological bases of the information revolution are less important to us than are the broad social impacts of these technologies. In this introductory chapter, we examine some of the perspectives offered about such impacts.

TOPICAL OUTLINE

Technology: Friend or Foe?
Doomsday versus Nirvana
Enter the Information Age
Personal and Professional Challenges

The Study of Communication
Varying Perspectives
Again: Friend or Foe?

TECHNOLOGY: FRIEND OR FOE?

Doomsday versus Nirvana

Technology is on the minds of many in our age. By some, it is considered to be the ultimate curse upon civilization. As if the problems of diminishing natural resources, pollution, and worker displacement were not enough, technology has presented us with the grim power to destroy life on earth but cannot resolve our uncertainty about how to prevent this catastrophe. But we have always had optimists, too. Technology to them has provided us with uncounted advantages: prolonging our lives, moving us easily from place to place, extending our physical and intellectual capabilities, and taking us into space.

So, too, has technology extended our abilities to communicate—so much so, in fact, that the evolution of human communication is at the heart of the social evolution of our species. Early tribal societies depended mostly upon speech for communication, the ancient civilizations added writing, the Renaissance was rooted in the spread of literacy and the industrial revolution in the growth of printing. Modern society rests upon organizational and public communication, and today we hear that computing and telecommunications are ushering us into the *information age.*

The acceleration of technological development is a major topic of our times. Whether our ability to apply technologies for beneficial purposes can match our ability to develop the technologies themselves has been the subject of such treatises as Peter Drucker's *Age of Discontinuity* (1969) and Simon Ramo's *Century of Mismatch* (1970). This challenge has also been interpreted in such large-scale social forecasts as Daniel Bell's theoretical *The Coming of Post-Industrial Society* (1976) and Alvin Toffler's popularly written *The Third Wave* (1981). In most such forecasts there is the socioeconomic theme of societies evolving from an industrial age to a postindustrial one, that is, shifting from manufacturing to high-technology and knowledge industries. Steel-making and automobile manufacturing are examples of the industrial economy; computers and telecommunications are postindustrial businesses. Yet as we shall see, the character of the information age (if this be an accurate label) is more complex than the replacement of one type of industry by another.

Enter the Information Age

If we are evolving toward a postindustrial society, an information one, or a communication age, it is the technological advances in computing and communication that are seen as driving forces. Just as the harnessing of energy to perform work was the mark of the industrial revolution, our growing technological power over information technologies may be the hallmark of the new age. This is a thesis found variously in such works as Marc Poratt's analysis of changes in the workplace (*The Information Economy,* 1977), Joseph Pelton's examination of world telecommunications (*Global Talk,* 1981), James Martin's consideration of the social impacts of telecommunications technologies (*The Wired Society,* 1978), Wilson Dizard's evaluation of world information orders (*The Coming Information Age,* 1982), or my own broad view of communication and change as described in *The Communications Revolution* (1982).

At least one consensus visible in many such treatises is that the communication technologies that are now becoming so common—e.g., cable television, video-cassette machines, new telephones, office technologies, personal computers, and **microprocessors** for automobiles, microwave ovens, and washing machines—are having increasing impact upon the social, economic, and cultural aspects of our existence. Less visible but increasing in impact on our everyday lives are supercomputers, **artificial intelligence,** telecommunications networks, and communications satellites. Although these are not the impacts that directly feed the broad doomsday or nirvana theses, they are nevertheless ones very much affecting our everyday lives.

The nature of work is changing as many types of employment involve more information work. Very few modern managers can stay competitive without acquiring mastery of the concept of office automation. We see examples where home shopping and banking services—**telemarketing**—and working at home—**telecommuting**—substitute communication for transportation. Health care is changing as new diagnostic machines are implemented and computers are used in the assemblage of medical information. Statistics show that the majority of our leisure is spent being entertained electronically (television still leads all forms). Our political system is affected because candidates who can command computer-planning resources and electronic-media campaigns stand the best chance of being elected. The question is even raised whether computers can improve our instructional practices, and our schools are scrutinized as we wonder whether our children are learning to cope with the new age.

Personal and Professional Challenges

The practical implication of the foregoing is that we are all challenged, in both our professional and personal lives, to make decisions about the new communication and computing technologies. Most of us wish to make the best of these technologies. We do not want to miss opportunities. We, too, want to get ahead. Least of all do we want to become the victims of these technologies. These desires raise in us such practical questions as: What can we expect from office automation, from new forms of home entertainment, from the so-called communication-transportation tradeoff, from computers in the schools, and from communication networks as new information utilities? There are broad theoretical questions: What long-range consequences will information technologies have on society? How can we understand their "fit" into our usual behaviors? What are their hazards and new opportunities? What is ahead in the mix of communication technologies and behavior, and what can we do to influence it? How can we maximize the long-range human benefits?

Much of the material in *Technology and Communication Behavior* was assembled for the purpose of not only framing such questions but also attempting to ferret out strategies for answering them. Although the record of our society in the application of communication technologies is not altogether commendable (e.g., our notable waste of the miracle of television and the invasion of our privacy by computers), we are learning much about the relation between communication technologies and behavior. Moreover, we are learning useful generalizations about cycles of planning, implementation, and evaluation. A fundamental assumption of this volume is that we can draw much of value from the social and information sciences for defining problems of technological adaptation and for formulating strategies for change. This is the practical goal of this book.

But at the same time, we should not overlook the larger issues. Indeed, another assumption of this book is that our success on the larger theoretical and philosophical levels of a rapidly evolving technological society will depend in part upon our success in responding to the daily details of change. An improved knowledge of the relations between communication technologies and human communication behaviors should aid us in shaping our longer range future.

THE STUDY OF COMMUNICATION

Varying Perspectives

If you have not studied communication previously, suffice it to say that it has been a serious topic of study for most of human intellectual history. Evidence indicates that even the ancient Egyptians had an interest in effective communication, as later did the Greeks (including Plato and Aristotle), the Romans, Western ecclesiastic writers of the Middle Ages, and Renaissance philosophers. Although this heritage has had its influences upon contemporary approaches to communication study, the modern behavioral sciences' approach to communication is mainly rooted in the latter half of this century, or to be more exact, in the time since the early 1940s. These modern origins initially reflected extensions of political science, social psychological, or sociological research in which communication phenomena were necessary parts of the main inquiry. As research into media, message, or effects phenomena progressed, human communication as a theoretical topic in itself grew, and the social-scientific approach has been one methodological rallying point.

Although the social-scientific approach is the bias of this volume, we should note that there are other recognized and valuable lines of communication inquiry. For example, there are the classically influenced *rhetorical* and *dramatistic* analyses of communication. There is the *critical* approach, which, among other things, stresses the sociopolitical evaluation of communication practices and institutions. Even if one is agreeable to calling communication a discipline (and many are not), it is not a unitary or even well-behaved one. The terms and theory of one approach may be inexact, if not inapplicable, in another. So beware.

Again: Friend or Foe?

Technological innovations have greatly accelerated in our times. We have so many alternatives now available to us, and those alternatives are so increasingly interconnected, that a new environment is forming. This environment can be likened to a *grid,* an omnipresent availability of communication links and services.

One social consequence of life on the grid is a quantum increase in what we might call *connectivity* in our modern communication age. This refers to the expanding ability of individuals and organizations to connect to a wide variety of information, communication, and computing services, as well as to each other. We are not so alone as before: The world is increasingly in touch.

We have new flexibility in our use of *space* and in our *mobility* because the grid often allows us to carry out our activities (i.e., working at a computer terminal) wherever we can locate the necessary equipment to link us into the network. We can telephone from our automobiles, boats, and airplanes, and, before this century is over, from our vest-pocket devices.

The grid also offers us new flexibility in our use of time. A video-cassette machine allows us to record our favorite programs and then view them at more convenient times, allowing us to fit many forms of leisure, work, and services (education, health) into our personal schedule. This practice is called *time-shifting.*

Our ability as individuals to adapt to life on the grid may be as critical to satisfaction—even survival—as was our earlier ability to shift from a tribal to a village society, or from a rural to an industrial life style. The adaptation of our communication behaviors to the potentials of communication technologies may be the mark of our age. Ultimately, we as individuals are the most important component in the information revolution.

Whether we are masters or victims of the new communication technologies depends upon our ability as a group to employ them wisely for human benefit. Whether they be friend or foe is ultimately the challenge that the new technologies pose to all of us, as workers, managers, consumers, or simply human beings in search of security and happiness, if not inspiration. This challenge, more than any other reason, is the fundamental rationale for learning more about the relation between the new technologies and our behaviors with them.

Topics for Research or Discussion

━━━━━ In a very broad view, it is possible to see relationships between major advances in human communication and advances in the structure of society. For example, ancient civilizations needed to keep records, which depended upon the development of writing. The Renaissance was associated with the development and widespread use of the printing press. The industrial revolution was facilitated by the mechanized press. Prepare a brief paper or discussion on some of the changes that computing and telecommunications technologies may bring to society. What types of changes do you envisage for government, work, or warfare?

━━━━━━━ Not all forecasts for the information society are optimistic. Prepare a report on the negative aspects of such changes. Examples include the invasion of privacy, glutting of broadcasting with trivial or violent content, de-skilling of workers, and undesirable urban growth due to industrial development. (Tip: See John Wicklein's book in the chapter references.)

━━━━━━━ Daniel Bell's concept of socioeconomic change *(The Coming of Post-Industrial Society)* has strongly influenced modern concepts of the growth of an information society. Examine Bell's book; summarize the essence of his thesis, then present your view of his position.

━━━━━━━ Consider how changes in computing and telecommunications may affect your professional opportunities. Prepare a *career impact* report on this topic. Be sure to contrast advantages and disadvantages of change. Finally, are there any actions that you should be taking now to prepare for your career?

━━━━━━━ That technology often exceeds our ability to implement it wisely is a thesis long held by a number of critics (e.g., see Ramo, Ellul, Wicklein, and Servan-Schreiber in the references). Examine the work of one of these critics (or others), then develop your own position paper on the issue. Given your position, what remedies would you advocate that we undertake to avoid negative consequences of our technologies?

References and Further Readings

Bell, D. *The Coming of Post-Industrial Society*. New York: Basic Books, 1976.

Branscomb, L. "Information: The Ultimate Frontier." *Science* 203:143–147 (1979).

Didsbury, H.F., Jr., ed. *Communications and the Future*. Bethesda, Md.: World Future Society, 1982.

Dizard, W. *The Coming Information Age: An Overview of Technology, Economics and Politics*. New York: Annenberg/Longman, 1982.

Drucker, P. *The Age of Discontinuity*. New York: Harper & Row, 1969.

Ellul, J. *The Technological Society*. New York: Alfred A. Knopf, 1964.

Haigh, R.W., G. Gerbner, and R. Byrne, eds. *Communications in the Twenty-First Century*. New York: John Wiley, 1981.

Martin, J. *The Wired Society*. Englewood Cliffs, N.J.: Prentice-Hall, 1978.

Masuda, Y. *The Information Society*. Bethesda, Md.: World Future Society, 1982.

McLuhan, M. *The Gutenberg Galaxy*. Toronto: University of Toronto Press, 1962.

Mosco, V. and A. Herman. "Radical Social Theory and the Communications Revolution." In *Telecommunications Policy Research Conference Proceedings,* edited by J. Schement, F. Gutierrez, and M. Sirbu, Jr. New York: Praeger, 1982, 58–84.

Naisbitt, J. *Megatrends: Ten New Directions Transforming Our Lives.* New York: Warner Books, 1982.

Pelton, J. *Global Talk.* Brighton, Sussex, England: Harvester Press, 1981.

Porat, M. *The Information Economy: Definition and Measurement.* Washington, D.C.: U.S. Government Printing Office, 1977.

Ramo, S. *Century of Mismatch.* New York: David McKay, 1970.

Robinson, G.O., ed. *Communications for Tomorrow: Policy Perspectives for the 1980s.* New York: Praeger, 1978.

Servan-Schreiber, J. *The World Challenge.* New York: Simon & Schuster, 1981.

Slack, J. and F. Fejes, eds. *The Ideology of the Information Society.* Norwood, N.J.: Ablex, 1984.

Teich, A.H., ed. *Technology and Man's Future,* 2d ed. New York: St. Martin's Press, 1977.

Toffler, A. *The Third Wave.* New York: Bantam Books, 1981.

U.S. Government. *The Global 2000 Report to the President: Entering the Twenty-first Century.* Vol. 1. Washington, D.C.: Government Printing Office, 1980.

Wicklein, J. *Electronic Nightmare: The New Communications and Freedom.* New York: Viking Press, 1981.

Williams, F. *The Communications Revolution.* New York: New American Library, 1983.

Williams, F. and R.E. Rice. "Communication Research and the New Media Technologies." Vol. 7, *Communication Yearbook,* edited by R. Bostrom. Beverly Hills, Calif.: Sage, 1983, 200–224.

Communication Processes and Contexts

The study of the technologies of communications can benefit greatly from understanding distinguishing features of the communication process. In this chapter we examine these features from a social-scientific and process view.

TOPICAL OUTLINE

On Analyzing Communication as a Process
The Fundamental Process
Communication Sources and Receivers
The Necessity for Coding
Overlapping Experience
Channel/Medium
Feedback and Interaction
Filtering
Specific Applications

The Contexts of Communication

Technological Applications

ON ANALYZING COMMUNICATION AS A PROCESS

The Fundamental Process

Most modern descriptions of human communication stress process: Communication is depicted as an ongoing series of events or operations that link humans and underlie their social structures. At the heart of this process are the fundamental steps required for message exchange. A general model of them is summarized in Figure 2.1.

Beyond the essential steps of the communication process, the model illustrates that this process involves quite different phenomena. For example, *intention* is essentially a mental or emotional state, whereas *creation* involves the cognitive and motor skills necessary to the overt behaviors of speech, writing, or other form of message emission. The *transmission-transportation* component is a physical process external to the human. It represents the necessary physical transformation of the message into a form that can bridge the gap between sender and receiver. In various definitions of the communication process this is referred to as the *medium* or *channel*. Like the behaviors of message initiation, message *reception* is a combination of physical (in this case, sensory) and cognitive behaviors, and *reaction* is initially a mental or emotional state.

Most examples of human communication, whether individual or group, organizational or public, can be summarized in a general form by the model shown in Figure 2.1. But a caveat is that there are many communication models that differ in various ways from the one presented here, and from each other. Most emphasize certain features of the communication process. In our present approach, we are not advancing a new model or even the use of models; the model is only for presentational convenience.

Communication Sources and Receivers

Basic to the analysis shown in Figure 2.1 is the involvement of human participants. *Source* and *destination* can represent individuals, groups, organizations, nations, or cultures. Often the exchange is in the context of a larger system. For example, my act of writing this page and your act of reading it are in the larger context of your studying the subject matter, attending a course, or reading in general.

The traditional media technologies of communications—print, wire, broadcasting, and photography—have allowed us broad choices among opportunities for extending our basic uses of sight and sound for everyday communication; now the new technologies are multiplying the choices. (Bear in mind that for most of human history, communication rarely went beyond the sound of our voices.) Today many of the new extensions of media fit under the broad term *telecommunications*. There is also the modern computer, one distinctive feature of which is that it can act as a communicating entity; for

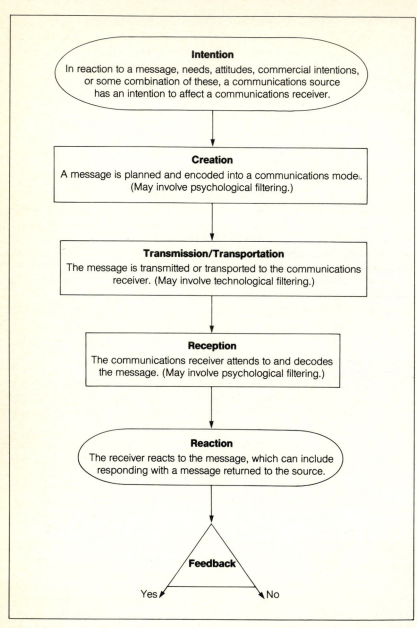

Figure 2.1 *General Model of the Communication Process*

example, a computer program that offers a variety of response options given human input can be thought of as communicating to us. In contrast, other communications technologies mainly involve storage or transmission of our messages.

The Necessity for Coding

Despite alledged evidence for the existence of extrasensory perception (which remains flimsy), a message from one communicating entity to another must be converted into a form that can be physically transmitted or transported. This requires *coding* by some system of symbolization. Coding may range from a simple gesture of the hand to a complex mathematical formula, but it is always required. This is one of the axioms of the theory of the communication process.

Systems for coding may use symbols that share some of the features of their referents (as in a drawing of a cat), called *iconicity,* or symbols that share none (as in the spoken word *cat*). Photography, line-drawing, and certain hand gestures are highly iconic, but most language is not.

All coding must be appropriate to the physical properties of the communication channel (discussed more later), and the communicating entities must share a certain minimal knowledge of a code. Highly iconic codes, such as international traffic signs and markings on shipping containers, are widely understood because people of various cultures share experience of the depicted referents. Codes with low iconicity are another matter; you must have experience with the specific code.

The development of a code for complex communication between humans and machines that is compatible with both has been a technological milestone. The great difference between our human "natural" language and the machine-level (binary) language of a computer was part of the challenge. Among the breakthroughs that made computers useful to nonspecialists was the development of go-between coding or translation systems whereby English or decimal math-like commands could be translated into the binary commands that computers could understand. The first notable example of this type of code was the programming language FORTRAN, which is short for *formula translation.*

Binary or digital coding offers many opportunities in modern technologically based communications. We shall see it frequently in subsequent chapters.

Overlapping Experience

To communicate we must be able to make certain practical assumptions about our intended receiver or audience. Imagine for a moment that you were asked to create a message for a galactic visitor. Could you make any assumptions about the visitor's use of language, hand signals, or sounds, or even about whether the visitor's sense organs were attuned to the frequency spectra of human communication by sight and sound?

Such questions should cause you to reflect upon the many assumptions that we necessarily make when creating messages. Even in everyday contacts with strangers, we make assumptions about language, motivation, likely responses, and the attitudes of our audience toward us. In a public media campaign, assumptions must be made in media selection, audience knowledge and attitudes, and persuasive strategies.

Our communication of new information is based upon an assumption of shared knowledge of other information. This textbook, for example, attempts to introduce new concepts and terminology in terms of concepts and vocabulary assumed to be already shared with you, the reader. To return to the idea of extraterrestrial communication, the plaque carried into deep space by Pioneer X contained, among other images, a depiction of the planets of our solar system in their relative sizes and order (but not distances) from our sun. An arrow that originates at the third planet from the sun points to a line drawing of the space satellite that contains the plaque. Presumably, a civilization advanced enough to try to interpret that plaque might also share the experience with us of the configuration of our solar system, thus providing at least one basis for reference in communication.

Channel/Medium

In communication research, *channel* refers to the method of message transfer between the source and destination: print, cable TV, broadcast, or in specific terms, for example, a given voice circuit. *Medium* is often used synonymously with *channel,* but if not, it usually refers to the physical basis for message transfer: sound, wire, visual, or the like. (In this text, we use *channel* for more general references to message transfer and *medium* to the physical form of transfer.)

As was noted in Figure 2.1, a message might travel between communicating entities by various means. However, there is a fundamental distinction between messages that are directly transmitted (such as speech, wire, or broadcast) and messages that are transported (such as printed materials or photographs). Of course, there are many combinations of the two forms. A television program can be recorded onto tape and the tape transported to a broadcast facility from which the message is transmitted.

The transmission forms of communication channels are closely associated with the media of the human effector (e.g., speech, gesturing, writing, drawing, or typing) and receptor (hearing and seeing) capabilities. Of primary importance in the study of communications technologies is the knowledge that they are essentially extensions of traditional communication channels in space and time. For example, broadcasting transmits speech far beyond the range of the human voice, and print (and its transportation) allows not only communication across great distances but also its preservation in time.

Feedback and Interaction

Like many communicating entities, you and I do not typically act only as a source or receiver of messages but are often engaged in both activities. That is to say, our competencies in communication are often equally applied to creating messages as well as to understanding them (including understanding what we ourselves communicate).

Feedback refers to secondary types of communications that we anticipate and interpret as responses to a primary message. For example, as you explain directions over a telephone, you expect the receiving party occasionally to say "ok," "yes," "uh huh," or some such acknowledgment. In a much larger context, feedback could refer to *letters to the editor* that are stimulated by a newspaper article.

In contrast, *interaction* connotes an active, more or less co-equal message exchange. Our most fundamental form of human communication, everyday speech, is often markedly interactive. We seldom speak to another human without the anticipation of hearing something in return. The more important forms of two-person interactions involve these active exchanges or transactions. An important challenge to communication technologies is making interactive channels available.

The various channel technologies, of course, differ in their potential for interaction and feedback. In most public communications, such as traditional commercial broadcasting or newspapers, feedback is not easily or immediately possible and may have to be transmitted in some alternative channel. Or, as in the case of traditional postal mail, return messages may be in the same medium but after a substantial time delay; feedback is *asynchronous*. These are in contrast with two-way or interactive channels, such as telephones, two-way broadcast systems, or linked computers, by which communications can be exchanged immediately.

One technological feature of the newly evolving network information systems is that, unlike traditional mass information systems, they are interactive. Technological capabilities for interactive channels also increase the potential for highly *personalized* communication systems. Personalized communication is considered to be the most powerful form of human communication in terms of influence as well as the development of interpersonal relationships (an important topic in modern communication theory).

Filtering

Although *filtering* is not usually included in simple communication models, its effects are an important consideration in our approach. *Technological filtering* refers to the incapability of some types of communication channels to accommodate the full representation of a message. For example, a traditional telephone circuit denies us the exchange of visual information and parts of the frequency range of voice communication; a written message cannot convey the emotional nuances of the human voice. To achieve communication effectiveness, we must choose channels that convey the critical portions of our messages, or if they cannot, we must employ compensations. (How can your voice or words convey a smile over the telephone?)

Psychological filtering is more complex; it broadly refers to the effects of attitudinal biases and code misinterpretation in the exchange of messages. For example, we sometimes try to lighten bad news. Often our attitudes bias what we perceive in a message; we look for what interests us. Also, there may be problems in understanding a code. Another bias is that we may expect only certain types of messages from a given channel—e.g., television is frivolous, or a computer printout is accurate.

Specific Applications

Simplified models, of course, have their limitations. The foregoing discussion is meant only as an introduction to a number of the basic suppositions about the communication process. The simplicity of the model that was shown in Figure 2.1 could be misleading because there are other characteristics in most communication situations that often loom in importance in research or practical application. Communication takes place in social contexts—e.g., conversations, groups, organizations, or broad public environments—and these have profound effects upon the process. Also, most communication is motivated, and this, rather than the general characteristics of the process, governs outcomes. Finally, in free enterprise societies, communication is a business. Whether a newspaper is published or a broadcast station stays on the air is determined by financial investment and the promise of profit.

In all, the basic components of our generalized model are only a starting point for analysis.

THE CONTEXTS OF COMMUNICATION

Having conceptulized communication as a process, we shall find it useful to consider next the broad distinctions among the social contexts of communication behaviors. These social contexts are contrasted in terms of the role of the individual, whether oriented as a private person in a variety of social roles, as a functionary in an organizational role, or as a relatively anonymous member of some public. The consideration of these contextual differences is, in part, an introduction to the distinctions made in the social-psychological literature of communications; but more practically, the social contexts constitute an intermediate step between the more abstract views of communication technology and communication behavior and the examination of the specific environments of technological applications.

We will concentrate upon three major types of contexts, bearing in mind that they may often overlap in everyday life.

1. INDIVIDUAL AND GROUP CONTEXTS: Mostly personal communications about self or self-with-others; communications that emphasize individual identity, including:

   A. *Intrapersonal:* communication with one's self, i.e., *within one*.

   B. *Interpersonal:* ongoing transactions between two individuals; a one-to-one context.

Table 2.1 *Social Contexts of Communication*

Context	Description
INDIVIDUAL AND GROUP	
Intrapersonal	Communication with yourself, as in thinking, monitoring your own speech and gestures, or writing as feedback.
Interpersonal	*Face-to-face:* Conversation between two people who are in physical proximity to each other.
	Point-to-point: Message exchange between two people who are not physically proximate to one another (e.g., over the telephone).
Small Group	*Face-to-face:* Communication among people who are physically proximate to one another and sufficiently few in number so that all can interact.
	Point-to-point: Message exchange in a medium (e.g., a telephone conference) that allows small-group communication among people not physically proximate.
Large Group	Communication to a physically proximate audience, as in a public speech, a performance, or a motion picture.
ORGANIZATIONAL	Communication that is necessary for the operation of an institution or a business; office communication. Could include the organization itself as a communications entity.
PUBLIC	Communication that is widely disseminated from a centralized source to individuals in large social configurations.

C. *Group:* interaction among a small enough number of individuals so that all can participate (small-group communication); gatherings at which individuals respond to a central communicator as an audience rather than interacting among themselves (group could be large).

2. ORGANIZATIONAL: Mixtures of the foregoing basic social contexts that support the operation of some type of institution, such as a business. The identities or roles of individuals are defined relative to the operation of the organization.

3. PUBLIC: Messages intended for very large groups of individuals; often the identity of the individual is subordinate to the identification of *audiences,*

publics, or *markets.* This category includes the traditional definitions of mass communication.

These distinctions can be summarized in terms of social units or structures, as shown in Table 2.1. Note, for example, that individual and group units incorporate such contexts as two-person or group interactions, as contrasted with the imposed structure of the organization or the lack of interpersonal links in public contexts. As we shall see in Part II of this volume, the implementation of technologies can be interpreted as serving these different structures.

TECHNOLOGICAL APPLICATIONS

Technological applications can be summarized conveniently as in Table 2.2.

Several generalizations can be made about the details in Table 2.2. First, all technologies are extensions of our natural human capabilities—that is, they extend our capabilities for visual and auditory communication. Technologies are not necessarily a new component in human communication: Our ancestors' carvings or paintings in caves, and even the first scratches in the dust, were a technology of sorts, a practical application of resources. Finally, we might well consider that technology has impacted the *how* of human communication much more than the *why.* We still communicate to relate to our fellow humans, to get things done, to instruct, and to seek or give pleasure. But our methods for achieving these ends are rapidly changing, perhaps more so in our time than any in the history of our species.

Topics for Research or Discussion

■■■■■■ Models of communication processes differ mostly for the convenience of the user. That is, they tend to highlight whatever the theorist or analyst wishes to emphasize. Prepare a report in which you develop your own general model of communication. Remember that it can be described in any of a variety of forms—e.g., verbal, diagrammatic, or numeric. Explain what you have emphasized.

■■■■■■ One can also use models for analysis of specific technological application. For this exercise, select an example of a contemporary communication technology—e.g., cable television, electronic mail network, computer timesharing, or electronic publishing—and prepare a systems type model of how the information is processed or distributed. See if the analysis will provide you with new insights or ideas about the technology.

■■■■■■ New media technologies make a much greater variety of options available to the public, a phenomenon some call *demassification.* Examine

Table 2.2 *Technological Applications*

Type of Communication	Traditional Technologies	Recent Innovations
Intrapersonal	Notes to self, diary, photographs, self-monitored feedback, calculators.	Audio or video tapes, computer programming, problem-solving with a computer.
Interpersonal	Point-to-point mail, telephone, telegraph, copying machines.	Facsimile, computer communications, electronic mail, mobile telephone, paging devices, personal videotaping.
Group (Point-to-Point)	Telephone conferences.	Teleconferencing with full audio and visual links, computer conferencing.
Large Group	Microphones, slide or overhead projectors, motion pictures.	Videoprojection, audience polling systems, personal audio.
Organizational	Memos, telephone, intercom.	Management information system, computer time-sharing, facsimile, teleconferencing, personal computing, word processing, electronic mail.
Public	Newspapers, radio, television, films, magazines, books, billboards.	Cable and pay TV, videotext and teletext, videotape, video or audio disk, portable radio and tape players (e.g., Sony Walkman), videogames, interactive TV, public information and computing networks, data radio.

several of the books on mass communication in the reference list for this chapter, then speculate on the consequences of demassification. Put into practical terms, what do you consider to be the likely future of mass media?

■■■■■■ In an era of technological implementation, we often find ourselves "doing the traditional nontraditionally." For example, we may modify our television viewing habits as new cable channels become available to us. Consider the types of change that may be going on in your life. Describe them. Then evaluate whether you think that these changes are positive or negative.

■■■■■■ Examine the details of Table 2.2 on technological applications to the various contexts of communication. What changes do you feel might be coming next in these applications? That is, how are our typical behavioral contexts of communication likely to be affected by still newer technologies? Evaluate these changes: Are they likely to be generally negative or positive?

References and Further Readings

Agee, W.K., P.H. Ault, and E. Emery. *Introduction to Mass Communications.* 7th ed. New York: Harper & Row, 1982.

Atwan, R., B. Orton, and W. Vesterman. *American Mass Media.* New York: Random House, 1982.

Barker, L.L. *Communication.* 2d ed. Englewood Cliffs, N.J.: Prentice-Hall, 1981.

Becker, S.L. *Discovering Mass Communication.* Glenview, Ill.: Scott, Foresman, 1983.

Berlo, D.K. *The Process of Communication.* New York: Holt, Rinehart & Winston, 1960.

Bittner, J.R. *Mass Communication: An Introduction.* 2d ed. Englewood Cliffs, N.J.: Prentice-Hall, 1980.

Davison, W.P., J. Boylan, and F.T.C. Yu. *Mass Media: Systems and Effects.* New York: Holt, Rinehart & Winston, 1976.

DeFleur, M. and S. Ball-Rokeach. *Theories of Mass Communication.* 4th ed. New York: Longman, 1982.

DeFleur, M.L. and E.E. Dennis. *Understanding Mass Communication.* Boston: Houghton Mifflin, 1981.

Dennis, E.E. *The Media Society: Evidence about Mass Communication in America.* Dubuque, Iowa: William C. Brown, 1978.

DeVito, J.A. *Communicology: An Introduction to the Study of Communication.* New York: Harper & Row, 1982.

Head, S.W. and C.H. Sterling. *Broadcasting in America: A Survey of Televi-*

sion, Radio, and New Technologies. 4th ed. Boston: Houghton Mifflin, 1982.

McCombs, M. and L.B. Becker. *Using Mass Communication Theory.* Englewood Cliffs, N.J.: Prentice-Hall, 1979.

Paisley, W. and R. Rice, eds. *Public Communication Campaigns.* Beverly Hills, Calif.: Sage, 1981.

Rogers, E.M. *Modernization Among Peasants: The Impact of Communication.* New York: Holt, Rinehart & Winston, 1969.

Rogers, E.M. and F.F. Shoemaker. *Communication of Innovation: A Cross-Cultural Approach.* New York: Free Press, 1971.

Sandman, P.M., D.M. Rubin, and D.B. Sachsman. *Media: An Introductory Analysis of American Mass Communication.* 3d ed. Englewood Cliffs, N.J.: Prentice-Hall, 1982.

Schramm, W. and W.E. Porter. *Men, Women, Messages, and Media.* 2d ed. New York: Harper & Row, 1982.

Sterling, C.H. and J.M. Kittross. *Stay Tuned: A Concise History of American Broadcasting.* Belmont, Calif.: Wadsworth, 1978.

Williams, F. *The Communications Revolution.* New York: New American Library, 1983.

The Technologies

Communication or *information* technology are the terms most frequently applied to the modern technologies of communication, especially those reflecting applications of computers, telecommunications, or their combination. In this chapter we survey the most common information technologies. Chapters 4 and 5 examine the processes of telecommunications and computing in greater detail.

TOPICAL OUTLINE

Perspectives on Technologies
  A Science of the Practical
  Looking at Technologies
  Looking at Applications

Selected Telecommunications Examples
  Satellites
  Optical Transmission
  Local Area Networks
  Cable Television
  High-Resolution Television
  Low-Powered Television
  Subscription Television
  Cellular Mobile Telephone

Teletext
Videotext

Examples of Recording and Playback Devices
    Video-Cassette Recorders
    Video-Disk Machines
    Compact Disks

Selected Computing Examples
    Microprocessors
    Personal Computers
    Supercomputers
    Artificial Intelligence

PERSPECTIVES ON TECHNOLOGIES

A Science of the Practical

There should be nothing mysterious, baffling, or even particularly new in the concept of technology in human communication. Throughout our history we have consistently expanded our communication capabilities by taking advantage of innovative tools or ideas, both of which are technologies in the larger sense. From the beginning of language and the scratching of territorial markings on rocks to the latest generation of communication satellites and supercomputers, we have always used a science of the practical to extend our transmission and reception of sights and sounds, to record ideas, and to facilitate our thought processes. Table 3.1 provides a brief chronology of advances in human communication that reminds us that the application of technology to communication is not a recent undertaking.

The majority of the technological innovations in human communication over the centuries have been advances in extending the range of our human effectors and receptors. Technologies like printing, the wired telegraph, photography, the telephone, broadcasting, coaxial cable, satellites, and fiber optics come to mind. Yet it is important to consider that technology is more than wires, equipment, or machines; it is also know-how. In this respect, the invention of language is a technology, as is the invention of writing, both of which were critical to the invention of printing. In fact, we should take care to remember that the know-how or "soft" side of technology is inextricably related to the equipment or "hardware" side. We often forget this because the equipment, by contrast, is so much more visible. The science of the practical is as much about knowledge and skill as it is about equipment.

Before departing from this brief historical reflection, we should make the point that all contemporary communication technologies share certain fundamental characteristics with the old. Like so many of the technological

Table 3.1 *Technological Advances in Human Communication*

35,000	B.C.	Cro-Magnon period; speculation that language existed
22,000	B.C.	Prehistoric cave paintings
4000	B.C.	Sumerian writing on clay tablets
3000	B.C.	Early Egyptian hieroglyphics
1800	B.C.	Phoenician alphabet
600	B.C.	Earliest Latin inscriptions
450	B.C.	Carrier pigeons used by the Greeks
130	B.C.	Library of Alexandria built
350	A.D.	Books replace scrolls
600	A.D.	Book printing in China
676	A.D.	Paper and ink used by Arabs and Persians
1200	A.D.	Paper and ink art in Europe
1453	A.D.	Gutenberg Bible printed
1562	A.D.	First monthly newspaper, in Italy
1594	A.D.	First magazine, in Germany
1639	A.D.	First printing press in North America
1642	A.D.	Early adding machine developed by Blaise Pascal
1709	A.D.	Copyright law in England
1791	A.D.	First Amendment to the U.S. Constitution
1819	A.D.	Flat-bed press invented by David Napier
1827	A.D.	Photographs on metal plates
1830	A.D.	Koenig steam press invented
1834	A.D.	Analytic engine (computer) principles; Babbage
1835	A.D.	Samuel Morse introduces the telegraph
1846	A.D.	High-speed printing
1855	A.D.	Printing telegraph; David Hughes
1866	A.D.	Transatlantic cable completed
1876	A.D.	Telephone invented; Alexander Graham Bell
1888	A.D.	Radio waves identified
1894	A.D.	Edison invents peep-show kinetoscope
1895	A.D.	Radio telegraphy; Guglielmo Marconi
1895	A.D.	Motion picture camera; Auguste and Louis Lumiere
1900	A.D.	Speech transmitted via radio waves
1906	A.D.	DeForest develops the Audion tube
1907	A.D.	DeForest Radio Telephone Company begins broadcasts
1912	A.D.	Motion pictures a big business
1915	A.D.	AT&T long-distance service reaches San Francisco
1920	A.D.	Home television speculated upon
1923	A.D.	Zworykin demonstrates partly electronic television
1927	A.D.	American Telephone and Telegraph Co. demonstrates TV
1927	A.D.	*Jazz Singer,* sound motion picture
1932	A.D.	NBC starts television station in Empire State Bldg.
1933	A.D.	FM radio demonstrated for RCA executives
1936	A.D.	*Life* magazine founded
1939	A.D.	Television broadcasting on a commercial basis
1942	A.D.	First electronic computer in U.S.

Table 3.1 *(continued)*

1943	A.D.	Wire recorders used in World War II
1946	A.D.	Xerography invented; Chester Carlson
1946	A.D.	Color television demonstrated by CBS and NBC
1947	A.D.	Transistor invented; Bell Laboratories
1949	A.D.	First stored-program computer
1951	A.D.	Color TV introduced in U.S.
1957	A.D.	Russia launches first earth satellite, Sputnik
1958	A.D.	Stereophonic recordings in use
1961	A.D.	Push-button telephones introduced
1962	A.D.	Telstar satellite launched by U.S.
1965	A.D.	Early Bird satellite launched
1968	A.D.	Portable video recorders introduced
1969	A.D.	Moon televised from Apollo flights
1970	A.D.	Microelectronic chips coming into wide use
1975	A.D.	Flat-wall TV screen invented
1975	A.D.	Home Box Office starts service
1975	A.D.	Fiber optic transmission highly developed
1976	A.D.	First wide marketing of TV computer games
1977	A.D.	AT&T tests fiber optic transmission
1977	A.D.	Qube interactive cable TV starts in Columbus, Ohio
1978	A.D.	Video-disk system test marketed
1978	A.D.	Public Broadcasting System satellite distribution
1979	A.D.	3-D TV demonstrated
1979	A.D.	U.S. Postal Service experiments with electronic mail
1980	A.D.	Home computer available for less than $500
1980	A.D.	New breakthroughs in space photography
1981	A.D.	Two video-disk systems widely marketed
1981	A.D.	Space shuttle Columbia has successful mission
1982	A.D.	European consortium launches multiple satellites
1982	A.D.	Advances in implementation of digital telephone
1983	A.D.	Integrated business programs for personal computers
1985	A.D.	Cellular mobile telephone widely marketed
1986	A.D.	Compact disk text publishing initiated

advances throughout history, they tend to build upon one another. Indeed, one main mark of our times is that we are witnessing many of the benefits of the combination of two relatively recent technologies: computing and telecommunications. These advances are especially important to communication behavior.

No classification scheme is entirely adequate when attempting to survey the various information technologies or their applications. But for present purposes—if just to have a scheme for keeping track of contemporary developments—it is useful to delineate the major technologies and applications environments as we do in the remaining major sections of this chapter.

Looking at Technologies

In the largest view, we typically consider various forms of telecommunications, computing, or combinations of the two. As for telecommunications, it is helpful to distinguish between technologies that represent specific types of transmission systems (e.g., fiber optics) and those that reflect a more general application, or even a business or industry (e.g., television). Also there are various types of sending, receiving, and recording devices (e.g., video-cassette machines) that are usually considered entities in themselves rather than aspects of telecommunications or computing. Table 3.2 illustrates a practical grouping of information technologies that is meant more as an illustration of the breadth of the subject matter than as a classification scheme.

Looking at Applications

There are many bases for classifying applications. It might be done by *level* of human communication (e.g., interpersonal, group, or public). Or it might be by traditional functions such as informing or entertaining. The list presented in Table 3.3 mainly represents environments in which applications are typically made. It is a helpful organizing scheme for keeping track of the different functional applications of communication technologies.

SELECTED TELECOMMUNICATIONS EXAMPLES

Satellites

The communication satellite is essentially a broadcast relay station which, because of its position far above the earth, can disseminate signals over a wider area than a land-based station. Traditional long-range broadcasting bounces signals off the ionosphere, a process prone to disruption. It is theoretically possible for three giant satellites to broadcast signals to the earth's entire surface.

Originally, satellite communication was especially complicated because earth stations had to track a satellite as it orbited the earth. This problem was alleviated by the *geosynchronous* satellite which, at 22,300 miles into space in equatorial orbit, remains in a fixed position over a given point (**footprint**) of the earth's surface. This, plus the ability to disseminate increasingly powerful signals from satellites, has drastically reduced the cost of earth stations. The multimillion-dollar earth stations that were the pride of technologists only decades ago are now rivaled by the relatively inexpensive "dishes" available at local electronic stores.

An important recent advance in satellite technology is the ability to repair or adjust satellites through use of space shuttles. Heretofore, the investment in a satellite was totally lost if it did not go into a designated orbit, and its life was

Table 3.2 *A Bird's-Eye View of Information Technologies*

1 TELECOMMUNICATIONS

  1.1 Specific Types

    1.1.1 Broadcasting
    1.1.2 Cable
    1.1.3 Facsimile
    1.1.4 Fiber Optics
    1.1.5 Local Area Networks
    1.1.6 Microwave
    1.1.7 Satellite
    1.1.8 Wire

  1.2 General Areas

    1.2.1 Radio

      1.2.1.1 Commercial
      1.2.1.2 Military
      1.2.1.3 Paging
      1.2.1.4 Telephone

    1.2.2 Telephone

      1.2.2.1 Cellular
      1.2.2.2 Technologies
      1.2.2.3 Videotext

    1.2.3 Television

      1.2.3.1 Cable
      1.2.3.2 High-Resolution
      1.2.3.3 Low-Powered
      1.2.3.4 Satellite
      1.2.3.5 Subscription
      1.2.3.6 Teletext

    1.2.4 Recorders and Players

      1.2.4.1 Video Cassette
      1.2.4.2 Video Disk
      1.2.4.3 Compact Disk

2 COMPUTERS

  2.1 Artificial Intelligence
  2.2 Mainframe Computer
  2.3 Microchip
  2.4 Microcomputers
  2.5 Microprocesser
  2.6 Minicomputers
  2.7 Networks
  2.8 Robots
  2.9 Supercomputers

Table 3.3 *Applications Environments of Information Technologies*

Banking and finance
Building and land planning
Communication as a business
Education
Entertainment
Governance
Group interaction
Health care
Home services
Information work
Interpersonal interaction
Management
Military
National development
Policy and regulation
Public information
Socialization

---

usually limited to a decade or less. Now, presumably, the life of a satellite is limited only by our ability to service it.

Also, shuttle technology will be the basis for building satellites larger than can be launched by existing rocket technology. In the not-too-distant future, we can expect to see very large satellites constructed in space, a technology referred to as *communication platforms*. The larger the platform, the more **transponders** a satellite can accommodate and the greater the power that can be supported by these channels. The more powerful the satellite-transmitted signal the smaller the earth station can be. (It is quite likely that wristwatch-sized earth stations will be feasible in the not-too-distant future.)

Of course, communication satellites have been extensively used for scientific purposes and over the last twenty years have become an integral part of the telephone, broadcast, and data telecommunications networks. One recent application, brought about partly by telecommunications deregulation, is the use of satellites to distribute broadcast signals directly to households and businesses.

Satellite communication is also a controversial international topic. Third World countries complain that 10% of the world's nations are occupying 90% of the broadcast spectrum. The controversy is also reflected in the question of who owns the satellite "parking spaces" above national boundaries. Should satellite operators pay for the privilege of orbiting over other countries, let alone for broadcasting to them? Much of the future of satellites is dependent upon international cooperation.

Satellites also have a major role in the establishment of private communication systems. Already it is possible to lease transponder time and earth stations to set up private networks. Current technology allows the linkage of small computers to one another through the use of earth stations and relatively low-powered signal systems.

Optical Transmission

Any carrier wave that can be modulated can carry communications signals, and light waves are no exception. Optical transmission systems are a major part of our telecommunications future. One of the greatest advantages of fiber-optic transmission systems is that the cost of the basic material (silicon) is far less than that of copper wire. Further, a small strand of fiber has a large capacity for signal transmission.

There are many challenges in the development of efficient optical transmission systems. One is the development of specific types of channel systems for light-wave transmission. Lasers, for example, concentrate light waves in a highly focused, parallel beam. Optical fibers guide light energy with thin glass strands. Another challenge is in how to amplify optical transmissions so they do not attenuate (i.e., die out) over long distances. Finally, there is the need to integrate switching systems—electronic ones—within optical transmission networks.

Local Area Networks

A local area network emphasizes several qualities, one being that it connects some group, such as linking an office, the buildings in an industrial complex, or offices located in a particular city or region. Another feature is that this network is purchased or leased for the sole use of its owner or client and not a common carrier (although the owner may choose to sublet services on it). Many local area networks are *broad band* in that they are designed to accommodate voice, text, and image transmissions.

A trend in local area networks is the expansion of their range of operation, for example, to serve a business that is housed in different parts of a city. Further, when users of the local area network wish to connect to the long-distance telecommunications network, they can now circumvent the local common carrier. That is, if you install a voice and data telecommunications system in the offices of your business, it is possible to connect directly with the long-distance network instead of going through the local telephone company. This is called *bypass,* and it is a controversial issue in the deregulation of the telephone industry.

Master satellite antenna systems, sometimes called *teleports,* represent another form of bypass. In large-scale applications, the development of teleports is roughly analogous to city or regional investment in airports or seaports because they are attractive to industry. On a smaller scale, the antenna systems and related networks that are installed in a commercial building are assets that can be leased to tenants just like electrical services, water, power, and air conditioning. (Many real-estate investors and landlords are actively entering the telecommunications business.)

Cable Television

Cable television refers to a wire-based carrier (coaxial cable) that simultaneously delivers multiple programs from a cable station operator to subscriber households. Programs can be obtained from a variety of sources, including: produced in the system's studio, reproduced from a tape or film recording, or received from a terrestrial or communication satellite network. The operator can make arrangements to relay existing broadcasting signals as they are received from other stations, can purchase special services (Home Box Office, the Disney Channel, Cable News Network), or can purchase or lease recorded materials. Programs are transmitted simultaneously on multiple channels from the cable company's *head-end*. Dissemination then involves *trunk* (open-line) cables, *feeders* that go with utility lines past individual dwellings, and *drops* that connect households (subscribers).

The cable industry, which began as simple *community antennas TV* systems (CATV), has become highly competitive. Although the growth of cable was overestimated at first, it still remains an attractive area for long-range investment.

Much attention was given in the 1970s to the prospect of establishing interactive cable systems. The return channels in such systems were not typically audio-video but were digital information systems. The best known example of a public system of this type was Warner-Amex *Qube* in Columbus, Ohio. Through the use of a key pad, viewers could order special programs, participate in quiz shows, order services or products, and respond to surveys. However, Qube has been unprofitable, and the interactive services are scheduled for discontinuance.

Many alternative services are possible for interactive cable systems, including text information services and fire, health, and security alarms (by which the appropriate agency is automatically notified of an emergency). Of course, many such services could be available via the telephone network (some cable companies lease phone lines for them). This raises the question of when or if, in this era of deregulation, telephone and cable companies will directly vie for the same markets.

High-Resolution Television

Commercial television in the United States and some other countries operates at a 525-line resolution, meaning that the picture signal scans the screen in 525 sweeps. (European standards are 625.) There is public anticipation of substantially increased resolution, say, up to 2000 lines. Higher resolution would make large-screen television sets much more attractive to buyers, would be much more suitable for video projection in theaters, and would accommodate the typical 80-column format of a computer text display.

The economic challenge is that if a genuine 2000-line system were to be installed, all existing equipment—cameras, recorders, editors, transmitters, and receivers—would be instantly obsolete. The main factor inhibiting the development of high-resolution television is not technical but financial: Who is going to make the investment? High-resolution television also requires more bandwidth than conventional channels, sometimes by a factor of five (meaning that a cable or spectrum location for one high-resolution channel might take up the same space as five traditional channels).

In the attempt to find a suitable compromise between the desired development of high-resolution television and the astronomical cost of replacing all existing systems, researchers have applied psychological research methods to determine how different types of technological enhancements will cause a television picture to be perceived as "different" or "better" by viewers. One enhancement, for example, has been to *refresh* only those parts of each frame that are different from the preceding frame. By transmitting only the *changed* portion of the picture, less bandwidth is required for a higher-resolution overall picture. Another strategy has been to use electronic image-enhancement techniques that involve replicating nearby pixels (the dots that constitute the television image). Further, there is the potential to combine these various innovations in a digital form of the television signal so that the final picture, although less complex than a 2000-line image, is seen as substantially improved over present image quality. The most acceptable techniques will likely meet this criterion plus have the least drastic impact upon equipment obsolescence.

Low-Powered Television

Low-powered television was sanctioned in early 1982 by the Federal Communications Commission. Essentially, it allows the establishment of stations with a sufficiently low signal radius so that they will not interfere with one another. This ruling opened up many opportunities for investors or public groups to enter the television business. The range of low-powered stations is usually less than ten miles. Low-powered stations had existed before, but they were mainly used to extend the range of traditional broadcasts.

Low-powered stations can be established for a fraction of the cost of a major station; usually the figure is under $100,000. This makes them attractive for specialized broadcasters (e.g., ethnic or foreign language groups), for public interest organizations, and for investors who normally would not be involved in large-scale television operations. Currently it is difficult to forecast the growth of low-powered television because of the large backlog of license applications and the limited experience of those stations that have managed to get on the air. Today it appears that low-powered television will bring new signals to rural areas and more diverse programming to urban ones.

Subscription Television

The dissemination technology for subscription television is either broadcast or cable distribution. The key distinction from traditional broadcasting is in the method for metering payment and the necessity of preventing unauthorized reception of signals. The latter is usually accomplished by scrambling the television signal so that a special device is needed to decode the picture.

The introduction of subscription television services was delayed by the active lobbying of theater and broadcast station owners. However, late in the 1960s, this lobby weakened, and subscription television was sanctioned by the Federal Communications Commission. There have been restrictions in terms of limiting subscription stations where sufficient numbers of traditional stations are established. Also, subscription stations are required to offer a minimum number of hours of nonscrambled programming.

The methods for determining fees range from devices in the subscriber's set that record the programs viewed, to the use of the telephone to order and charge programs. Scrambling techniques have usually involved frequency or other electronic manipulations of the broadcast signal. The problem has been that electronics enthusiasts have usually been able to build decoders so they can receive programs free. There is a cost limit to the complexity of decoders, which has tended to make these devices easy to *crack* or emulate. The law has been vague, too, as to the use of decoders. In fact, a company in the Los Angeles area in the 1970s that sold unauthorized decoders called itself *Pirate Television*.

Cellular Mobile Telephone

For most of its existence, the mobile (radio) telephone has been severely limited in use by the number of available channels. Traditionally this service involved a single large radio transmitter-receiver serving a given area. The problem was that the number of telephone calls that could be handled was limited by the broadcast frequencies allocated to the service. Service was difficult to obtain, discouraging growth. *Cellular systems* bypass this problem by dividing a service area into a grid of *cells,* from each of which low-powered receiver–transmission systems can be accessed and connected to the wire network. Because the signals are low-power, frequencies can be used many times in a large metropolitan area. As a caller travels from one cell to another, computer-assisted switching systems transfer the call without interruption from the frequency of the old cell to that of the new one. This greatly increases the number of available channels.

The future of mobile telephones may extend far beyond the instruments located in automobiles. Cellular technology can bring telephone service to remote areas where the cost of a wired network would be prohibitive. Perhaps more intriguing, however, is the prospect of being able to carry a miniaturized mobile cellular phone on your person to receive or place calls anywhere you are. Although such applications overlap the radio paging market, the two may eventually merge, giving us both voice and data communications. As telephone service is increasingly identified with the individual person rather than a particular telephone, we may someday be assigned a personal telephone number that will follow us wherever we go.

Teletext

Teletext* typically refers to text services that are distributed via broadcast channels. These are services that have frequently been test-marketed by large cable systems or publishers. Sometimes the text is sent "piggy-back" in the vertical blanking interval (the empty space) between television screen images. Most require a special terminal to be connected to the subscriber's television set. As contrasted with text systems that use personal computers as terminals, special teletext terminals allow a much more sophisticated text design, graphics, and color. However, this advantage has been offset by the necessity for the vendor to lease or furnish expensive terminals that cannot be used for other purposes.

Most teletext is not fully interactive because the customer is not directly connected with the main database. Instead, a selected number of pages is broadcast, and a memory unit stores them for the customer's use. The customer, then, has access only to those pages until they are refreshed by another broadcast sequence. Among the advantages of teletext is that it does not require telephone lines or the sometimes complicated *log-on* procedures for linking one's computer to a videotext service. The major disadvantage found in market studies is that not enough potential subscribers have sufficient interest to pay the subscription fees (and terminal costs) for the services to be profitable. Accordingly, one possible future for teletext is to incorporate advertising. The challenge here, however, is whether text services will ever attract a sufficiently large subscriber base to be attractive to major advertisers.

Videotext

Videotext increasingly refers to a more computer-oriented form of public text system. Subscribers link their personal computers via telephone to a central computer system that offers a wide variety of text services. Videotext can be likened to computer time-sharing systems in which one accesses different databases. The best known public text systems are *The Source, CompuServe,* and *Dow Jones News/Retrieval.* The databases may contain news headlines, theatrical or travel schedules, classified ads, columns, news, or electronic mail, to name a few options.

Costs for using a videotext service typically include an initial membership fee, on-line time charges, and sometimes special surcharges for special databases. Although advertising is not usually a component of videotext, there are services that offer classified advertising. Experience with videotext indicates that continuing customers typically have special uses for information that the service can offer, such as for investment data, travel schedules, or electronic mail.

One major advantage of videotext is that the use of personal computers

---

*Sometimes *teletext* and *videotext* are used synonymously; the distinction made in this chapter will be found in most uses. Also, both names may be found with the final *t* omitted.

rather than expensive single-purpose terminals restrains service costs. Also, all a new subscriber (with the proper equipment) needs to initiate service is to call into it to register a subscription and get log-on information. There is no need to install an expensive terminal to one's television set. Also, services are increasingly easy to use as superior computer software is developed to support them.

EXAMPLES OF RECORDING AND PLAYBACK DEVICES

Video-Cassette Recorders

The capability of recording video images with sound has been available since its introduction by the Ampex Company in the mid-1950s. This technology was rapidly adopted by the broadcast industry and replaced the older kinescope, which transferred film images into television ones. Although small reel-to-reel, black-and-white videotape machines were available by the late 1960s, they were used mainly in noncommercial broadcasting, schools, or industrial training. By the mid-1970s, the cassette versions of the machines were introduced in a 3/4-inch tape format that immediately became the standard of the industry because of widespread adoption, including by the U.S. Government. Later in the 1970s, home versions of the machine were introduced, and by the mid-1980s video cassettes were the best-selling home-entertainment technology.

The basic operation of a video-cassette player involves the recording and subsequent sensing of electronic or magnetic patterns that are the analogs of broadcast signals. This is accomplished by recording and playback heads that spin while the tape is pulled across their surfaces. The width of the tape can be used to full advantage because it moves at a comparatively slow speed compared with earlier systems. Home video-cassette machines are made in two incompatible formats: Beta and VHS. The latter may eventually become the industry standard.

Research indicates that the most frequent use of video-cassette tapes is for recording programs to be watched later, at a more convenient time for the viewer, an application called time-shifting. Statistics in the late 1980s indicate that the playing of prerecorded cassettes (specially rented theatrical films) is also a major use of these machines. Many films are grossing as much profit, or more, from cassette sales and rentals as from theatrical showings.

Video-Disk Machines

Video-disk* machines were brought out with great fanfare in the early 1970s as manufacturers of machines and disks compared their product with the more

---

*Disk is also spelled *disc,* and the name can be found as one or two words—i.e., *videodisk* or *video disk.*

expensive cassette machines. The manufacturers believed that they could undersell the cassette machines and the tapes. Another presumed advantage was that, as compared with the mechanical requirements of a tape cassette, disks could be produced inexpensively (raw materials were approximately ten cents per disk). They were also supposedly indestructible. (They were not: A major problem was warping.)

More sophisticated *random access* disk machines do have a major advantage over cassette technology. Any program material on the disk can be addressed by a numerical code, making it possible to move at will among different areas of the disk rather than having to read in a linear sequence as is necessary with tape. It has been presumed that in addition to playback of traditional television program materials, the disk would be especially useful for storing pages of text, still images, or even data. Moreover, all these modes could be mixed on an individual disk and individually accessed. (A demonstration version of the Sears Roebuck catalog has been put on disk.) Together these advantages were the basis for imagining video disk as a new form of publishing technology.

The main disadvantage of the disk as a home consumer product was that, at the time, users could only play prerecorded material because the recording technology was expensive and not available for home use. The *playback-only* feature was attractive to motion picture companies because the disk could not be used to copy programs off the air or to make duplicates for distributions to friends. (This was one of the motives for the unsuccessful suit brought by Walt Disney Productions and MCA Universal in the late 1970s against the Sony Corporation; they claimed that Sony's video-cassette machines violated copyright laws.)

The disk machine failed in the early 1980s as a home-entertainment machine. Poor quality control of disks resulted in public disappointment with the product, and the lack of a home-recording capability made customers prefer tape. Added to this, the Japanese were able to export video-cassette machines for prices that were competitive with the RCA Company's $500 disk machine. When RCA wrote off millions and left the video-disk business, it sealed the fate of the home market, for this decade at least.

In the meantime, Japanese manufacturers have continued in the disk business. The most successful sales are of full-featured machines used for training purposes in government and industry. Machines are also appearing that have recording capabilities (although they are very expensive—some around $35,000 in the mid-1980s). On the other hand, in the home market, the compact audio disk seems to be taking the role that manufacturers had hoped for the video disk.

Compact Disks

The compact disk uses the same recording and play-back technology of the video disk but is smaller in size. It was first used mainly for audio recording but has recently become a promising medium for "read only memory" (thus the

"CD-ROM" name often used) when used in conjunction with computers. As of the late 1980s, the compact disk, unlike the video disk, as an audio recording product had gained a place in the home consumer market. Moreover, the use of CD-ROMs as a means for "publication" of catalogs, databases, and computer programs, or any application for computer retrieval, seemed to be growing in promise.

SELECTED COMPUTING EXAMPLES

Microprocessors

Microprocessor chips are integrated circuit devices etched by microlithography and other **large-scale integration (LSI)** techniques. Literally thousands of microelectronic components can be interconnected to form a central processing unit. A microprocessor can be a fundamental unit in a wide variety of microcomputer applications—in computers and also in many other applications where some type of "intelligent" circuit is required, such as pocket calculators, control mechanisms for robots, and telephone switching units.

A miniaturized central processor in a microcomputer performs the same logical operations and can control storage and input/output devices in the manner of its larger counterparts in mini and **mainframe** computer systems. Although its design is expensive, its ease of manufacture makes the microprocessor very inexpensive to reproduce. It requires very little power or cooling. Some microprocessors for personal computers wholesale for one or two dollars, an unbelievable figure only a decade ago.

New manufacturing techniques underlie the great advantages of microprocessors. One is the technology of **very large scale integration (VLSI),** which involves the ability to manipulate microscopic elements on a chip and to transfer a design from a planning chart through lithographic techniques to the actual etching of a highly miniaturized replica. These techniques have changed the process of manufacturing computer circuits from the physical processes of wiring to a photographic-like process.

Much of the history of the microprocessor has been visible in terms of the introduction and popularization of the microcomputer. But probably further reaching is the evolution of microprocessors to bring "intelligence" to many of our everyday devices, ranging from microwave ovens to fuel injection in our automobiles.

Personal Computers

In the early 1970s, with the advent of integrated circuits and the possibility of reducing the central processing unit of a computer to the surface of a single chip, designers began to see the possibility of adding the other components of a computer to a single board that could be housed in a desktop unit. At the time, many manufacturers saw this more as a curiosity than as a product to be refined and marketed. The first such machines, **microcomputers,** were offered to electronics or computer enthusiasts in kit form. However, at the same time, several visionaries saw more promising possibilities. Alan Kay, a scientist at Xerox, pondered the concept of a book-sized machine that could serve as an executive workstation and that could be operated by a simple selection of icons. Although Xerox never built the small machine, many of Kay's concepts later underlay the development of "lap" computers. His icons were pursued in Xerox's *Star* workstations but were eventually most widely encountered in Apple's popular *MacIntosh* computer.

Shortly after the mid-1970s, three companies successfully introduced inexpensive personal or home computers. The largest of these contenders was the Tandy Corporation, which brought the *Radio Shack* computer to market via its chain of retail outlets. Another was Commodore International, a calculator and chip manufacturer that entered European and American markets. The third was Apple Computer, founded by two college drop-outs. By the early 1980s, the success of these early entrants had encouraged many other companies to enter the competition, including International Business Machines, whose *PC* and its many available business-oriented programs eventually set the operating standard for that segment of the market. Shortly thereafter, and probably somewhat independent of IBM's success, the home-computer market (including videogames) began to deflate, and a number of major contenders were forced from this line of business.

The business market for personal computers continues to grow, both in numbers and sophistication of machines, including the interconnection of desktop and mainframe computers. The home market has not grown according to original expectations. And although many small computers have found their way into grade and high schools, they mostly populate special labs where computer literacy is taught, rather than being used as a fundamental technology for instruction.

Perhaps one of the most exciting growth areas for personal computers is the university market. There have been cooperative projects at several major universities to develop powerful personal workstations. These have large memory capacity, high-resolution screen images, and the ability to be connected in a variety of communication networks. We may first see the next generation of personal computers in this environment.

Supercomputers

The first supercomputers were built to help decipher military communications codes and design weapons. Today the U.S. National Security Agency and the Department of Energy nuclear weapons laboratories have the biggest collections of the world's supercomputing machines. They are vital to the U.S. military strategy of relying on qualitative military superiority to offset the Soviet numerical advantage.

Today's supercomputers are essentially advanced versions of ordinary computers, and the drive to improve their performance is encountering fundamental physical barriers. One after another, supercomputer makers are discarding the engineering principle that for the past 40 years has been the foundation of virtually all computers. As outlined in 1946 by mathematician John von Neumann, a computer's central processor gets its instructions and data from a main memory step-by-step, pausing after each step to send its results back to the memory. All this happens in what is in human terms blinding speed. But the frequent delays as the processor waits for the data to come and go are an eternity for a system that measures time in billionths of a second. In the past, engineers minimized the problem by making the transistors on the computer's integrated circuits smaller and packing them closer together. That reduces the distance the data must travel. But the more transistors on a chip, the more heat generated. Unless this heat is removed, the circuits will melt. To cool some contemporary supercomputers, the chips are immersed in a liquid refrigerant.

Departing from the von Neumann design, makers of supercomputers are turning to multiprocessor strategies. Instead of channeling information into what amounts to an electronic funnel, the latest supercomputers utilize several processors operating in parallel—a few machines have scores of them. Tomorrow's multiprocessor giants will routinely have thousands, even millions, of parallel processors. Assigning a large number of processors to tackle different parts of a job allows computers to attain fantastic speeds. Several contemporary computers have exceeded a million operations per second, and it might be possible to reach a trillion by the 1990s.

With machines 1000 times faster than today's supercomputers, one might think that the thirst for computing capacity will soon be satisfied. Only about 35 supercomputers were sold worldwide in 1984; most projections for the next decade call for annual shipments to increase to approximately 200 supercomputers or more. One of the major barriers to development of parallel processing is that it requires an entirely different type of software, since most existing programs are geared to serial programming. To exploit parallel processing, new programming techniques are needed. Some scientists are even calling for entirely new program languages. Others maintain that software cannot be developed until a programmer knows which parallel architecture will use the software. The programming method that works on a machine with its processors arranged in a branch-and-tree organization, for example, will not work with a computer that has a grid of parallel processors.

Some of the major and projected applications of supercomputers include

- *Meteorology:* predicting the movements of storms and detecting broad changes in atmospheric conditions and temperatures.
- *Aerodynamic and automobile design:* modeling the flow of air over a wing or the stress points of an automobile, making it possible to test engines without building prototypes.
- *Oil and geophysical explorations:* using sonic and other data to predict the characteristics of underground reservoirs and geologic faults.
- *Intelligence gathering:* sifting through millions of satellite and other transmissions, particularly data traffic, as part of routine electronic spying.

■ *Code deciphering:* testing millions of combinations in sophisticated efforts to crack codes.

■ *Nuclear energy research:* simulating the actions of atomic particles, both for weapons research and nuclear energy projects.

■ *Graphics and film animation:* creating three-dimensional models used in special effects work, particularly in science fiction movies and in medical scanning techniques.

■ *Automatic language translation:* having the capability to select among the many alternative interpretations when translating from one language to another, especially in using semantic evaluations for this process.

Artificial Intelligence

One of the most striking applications of modern computers is in a field long known as **artificial intelligence (AI).** Here the concept of intelligence refers to abilities such as learning, adapting, reasoning, guessing, and emulating. Further, this implies that a computer could have the ability to improve its own abilities.

For many years, artificial intelligence was an esoteric field of research mainly pursued for theoretical ends. However, in the 1980s, AI techniques began to be used for applications in which human expertise and reasoning power that had been "captured" by the computer could be applied to problem solving. Many of these applications are now being called a type of *knowledge engineering.* Systems of this type create solutions based upon knowledge derived from human experts and codified for use by the computer. Developed over the last decade, they have demonstrated an ability to perform many practical tasks. The first projects to emerge from expert-systems research have been so-called shells—the language and logic needed to draw conclusions from codified expertise. By using shells, expert-systems research can concentrate on capturing the expertise needed for certain types of problem solving, instead of having to develop the codification system every time. Most expert systems process knowledge as *if-then rules* in patterns characterized as *backward chaining.* The process begins with a hypothesis, determines rules that support the hypothesis, then searches the knowledge base for relevant relationships. In this manner a system of tested if-then rules can be applied to a practical problem. Thinking about medical diagnoses as an if-then process is a rough analogy here. Other practical applications of contemporary expert systems include

■ Interpreting geologic data
■ Predicting chemical processes
■ Planning complex projects
■ Training human experts
■ Monitoring satellites

- Designing telecommunications networks
- Designing factories

Topics for Research or Discussion

■■■■■■■ Go to your library and sample advertisements for video-cassette recorders at one-year periods from approximately 1978 (when they first became popular). Examine cost data from these sample years and plot it on a chart. Also make notes of the various special features offered on the machines. What generalizations can you make about the sales of video-cassette machines in your community (or the United States)? Have the prices of these machines bottomed out? What do you think the future holds for their design and sales?

■■■■■■■ Prepare a brief analysis of the marketing of personal computers. For example, trace advertisements at about six-month intervals in a popular newspaper dating back to about 1978. Examine the features of personal computers highlighted in these ads, especially size of memory, availability of disk drives, or other special features. Record information on the costs of the machines. What brands have come and gone? What are some of the major contrasts between the personal computers of today and those offered for sale in 1978 or 1979?

■■■■■■■ During the early 1980s, videogames became very popular, but then interest seemed to die out. How would you analyze that situation? Do you think that videogames were only a fad or that their designers or manufacturers were not sufficiently creative to keep up with market tastes? What future do you see for the videogame type of media applications? For that matter, do you see applications beyond the use for games?

■■■■■■■ It is sometimes joked that organizations in our society hold more personal information on us than we can find in our own home files. Make a list of all of the examples of this type of information (e.g., driver's license, health record, academic record, and the like). If all of this information could be assembled in one database, what overall generalizations might be drawn about you, your life style, your problems, or prediction of your behavior? Do you think there should be comprehensive laws affecting the availability of this information to others or the transmittal of it to a central organization? What problems or challenges do you see in this area?

■■■■■■■ For purposes of comparison, select the programming available for several representative days from over-the-air broadcast television stations as compared with those available via cable television service. From your analysis of types of programs, what are the major contrasts in content or function of the programming? Or put into practical terms, what, if any, alternatives for viewer satisfaction does one gain from subscription to cable service in your area? In the broadest picture, what social consequences do you see as a function of increased program choice?

References and Further Readings

Dordick, H. *Understanding Modern Telecommunication.* New York: McGraw-Hill, 1986.

Forester, T. ed. *The Information Technology Revolution.* Cambridge, Mass.: MIT Press, 1985.

Forester, T., ed. *The Microelectronics Revolution.* Cambridge, Mass.: MIT Press, 1981.

Martin, J. *The Wired Society.* Englewood Cliffs, N.J.: Prentice-Hall, 1978.

McHale, J. *The Changing Information Environment.* Boulder, Colo.: Westview Press, 1976.

Mosco, V. *Pushbutton Fantasies: Critical Perspectives on Videotex and Information Technology.* Norwood, N.J.: Ablex, 1982.

Parker, E. "The New Communication Media." In *Toward Century 21: Technology, Society and Human Values,* edited by C.S. Wallia. New York: Basic Books, 1970, 97–106.

Plude, F. "A Direct Satellite Delphi Study: What Do the Experts Predict?" *Satellite Communication* (April 1981): 32–36.

Rice, R.E. and Associates. *The New Media: Communication, Research, and Technology.* Beverly Hills, Calif.: Sage, 1984.

Singleton, Loy A. *Telecommunications in the Information Age: A Nontechnical Primer on the New Technologies.* Cambridge, Mass.: Ballinger, 1983.

II

TECHNOLOGICAL ADVANCES

Much as the machine enables us to extend our physical capabilities for labor, communication technologies extend our human abilities for sending, receiving, and processing information. At the heart of these technologies are advances in telecommunications and computing. The two chapters in this section describe more of the details of these topics.

4

Telecommunications

Telecommunication literally means *communication over a distance*.
Typically, the distance is greater than we can accomplish by our
natural capability for sight or sound communication. In this chapter
we examine some of the fundamental bases for the processes that
enable telecommunications.

TOPICAL OUTLINE

WHY EXAMINE TECHNOLOGICAL DETAILS?

There are, of course, many alternative approaches to the study of the technologies of communication. In the previous chapters, the emphasis was on how technology has been applied to facilitate fundamental process components of communication, the components emphasized in Chapter 2. The present chapter examines a class of technologies. Although to the nontechnical reader, the coverage may seem a bit detailed at first, it does pay dividends in terms of understanding the many practical and theoretical implications of our uses of telecommunications. In brief, this chapter stresses how technology extends, supplements, and amplifies our human capabilities for signal transmission.

GENERAL CHARACTERISTICS OF TELECOMMUNICATIONS

Communication Over a Distance

Again, *telecommunication* refers to any system or process by which message transmission takes place over a greater distance than we can accomplish by our more traditional means of face-to-face communication. Further, telecommunication implies an active transmission process (e.g., the telephone) rather than the transportation of messages (as with a letter). Although ancient systems for smoke, fire, or drum signals would fit this definition, as would later semaphore or light-signaling systems, modern telecommunication particularly refers to wire, broadcast, and more recently, optical transmission systems. It also reflects the growing convergence with computers.

Major Processes

We can refer to a generalized model of telecommunications (Figure 4.1) to remind us of the relative role of technological components in the communications process. In broadest terms, telecommunications components involve the following functions:

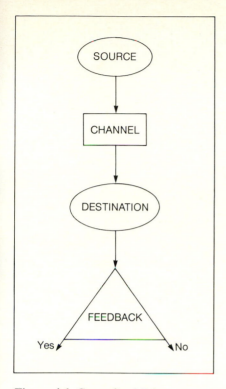

Figure 4.1 *Generalized Telecommunications Model*

1. A message source creates signals in a form codable and transmittable in the physical medium of a telecommunications channel.
2. The signals are transmitted via the telecommunications channel.
3. The signals are received in a form decodable by the message destination.
4. There may be a potential for feedback.

It is easy to think of examples of the foregoing processes that are involved in everyday life. For example, when you speak into a telephone mouthpiece, the acoustic patterns of your voice are converted into analogous electrical patterns and transmitted via a wire network to the listener. The receiver in the listener's handset then converts these electrical impulses back into the acoustic forms of intelligible speech. Or to look at the processes from the receiving end, when you watch a television show, the broadcast signals are converted into visual and auditory forms interpretable by you. They reached your television set by a broadcast transmission process initiated by a television station. The station's transmitter converted images and sounds into broadcast transmissions.

Table 4.1 *Common Telecommunications Systems*

System	Medium	Signal Transmission	Interactive	Switched
Telegraph	Wire	Digital	Yes	No
Telephone	Wire	Analog	Yes	Yes
Coaxial TV cable	Wire	Analog	No*	No
Optical	Optical fiber, laser	Analog or digital	Optional	Optional
Radio telegraph	Broadcast	Digital	Optional	No
Packet radio			Optional	Yes
Mobile telephone	Broadcast	Analog	Yes	Yes
Radio pager	Broadcast	Digital	Usually no	No
CB radio	Broadcast	Analog	Yes	No
Commercial radio	Broadcast	Analog	No	No
Commercial television	Broadcast	Analog	No	No
Microwave relays	Broadcast	Analog/digital	Optional No	Optional No
Communications satellites	Broadcast	Analog/digital	Optional	Optional

*A few are interactive, such as Warner/Amex's Qube in Columbus, Ohio.

Common Types of Telecommunications Systems

As much as telecommunications systems appear to differ from one another—as a television show differs from a data network—they all have the common components or features summarized in Table 4.1. These reflect the three basic processes described earlier. All systems involve the transfer of messages via the transmission of coded energy forms—i.e., as signals. There must be a component in the source that is capable of creating signals for a given type of channel. The channel must link source and destination. Finally, the destination must have the capability of receiving and decoding the signals into a message form. Two further characteristics are added to Table 4.1: whether a system is interactive and whether you can change the transmission paths or circuits by switching.

Let us turn to a discussion of these basic characteristics because they help us understand the fundamentals of telecommunications and the operation of different practical systems.

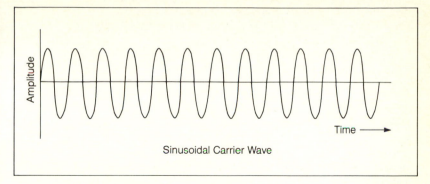

Figure 4.2 *Wave Properties*

THE MEDIA OF TELECOMMUNICATIONS

Carrier Waves

It is the transmission medium that is the *energy carrier* for messages in a telecommunications system. As can be seen in Table 4.1, forms of electrical, radio, and light energy are the bases for most telecommunications systems. All such energy forms have wave characteristics, which in simplest terms are comprised of pulses of energy that vary over time in their strength (amplitude) and frequency. Figure 4.2 illustrates these properties.

Frequency and Bandwidth

Frequency is the number of cycles per second, a cycle being one complete wave variation, measured from one point on the wave pattern until a similar point recurs. It is often expressed in units of 1000 hertz, or 1 kilohertz (kHz). There is a direct relation between the range of frequency variations in an analog type of message and the capacity of the communication medium to accommodate that range. This frequency range is called *bandwidth*. In simple terms, its measurement is the highest frequency minus the lowest. A telephone circuit, for example, requires a frequency range of about 4.0 kHz; an AM radio channel is assigned 7.4 kHz.

It is easy to think of examples of differences in bandwidths. For example, a voice-grade telephone circuit has just enough bandwidth to allow speech to be intelligible; it is insufficient for high-fidelity music. There are vast differences in bandwidth requirements among common types of telecommunications systems. For example, a television channel (broadcast on one wire of a coaxial cable) requires roughly 1000 times the bandwidth of a telephone circuit. (This is 4600 kHz, more often expressed as 4.6 megahertz [MHz], or millions of cycles per second). To visualize the magnitude of the difference between a telephone and a television channel, consider the contrast between the cross section of a 1/2-inch garden hose and that of a 30-foot subway tunnel.

Multiplexing

Obviously, it would be advantageous to be able to send more than one message at a time over a given communications channel. This common practice is called **multiplexing.** One method involves dividing a channel into segments of smaller bandwidth, each of which then serves as a separate channel. Another method is to allocate messages to patterns of sequencing in time (part of one message is sent, then part of another, etc.). In either case, transmission and receiver devices are able to separate and recombine messages; the resulting messages are virtually indistinguishable from the original messages by the users. Multiplexing allows much more efficient uses of communications channels, and techniques for it are still being improved.

Spectrum

As you probably know from your experience with the frequencies of different radio stations, broadcast channels are separated from one another by assigning them different frequency ranges for their operation. This is called *spectrum allocation,* and it is applied to all types of broadcast channels. Table 4.2 illustrates how a major part of the broadcast spectrum is divided.

Table 4.2 *Spectrum Allocation*

Designation	Frequency	Application
Low frequency	10^4	Long-wave radio formerly for transatlantic telephone
Medium frequency	10^5	AM broadcasting
High frequency	10^6	Short-wave radio
		Ship-to-shore radio
		Telephone
Very high frequency	10^7	FM broadcasting
		VHF television
Ultra high frequency	10^8	UHF television
Super high frequency	10^9	Microwave relays
		Satellite communications
Extremely high frequency	10^{10}	
	10^{11}	

Modulation

Whether we are talking about electrical current, radio, or light, the basic *linkage* of a telecommunication system depends upon a source being able to impose a particular signal pattern upon a carrier wave. Often this is a general form called *analog modulation,* in which the carrier signal is modified in a manner that has characteristics analogous to the original signal—e.g., speech sound-wave patterns converted into electrical wave patterns. Nearly any energy form that we can modulate has the potential to be a communication carrier (e.g., light, water flowing in a pipe, or vibrations in a steel rod). Two common types of analog modulation are *amplitude modulation* and *frequency modulation.* As the terms denote, amplitude modulation imposes variations in amplitude upon a carrier wave, and frequency modulation imposes variations in frequency. (Yes, these are the same as the distinction between AM and FM radio.)

Pulse Code Modulation

Modern pulse code modulation divides a waveform into a selected number of discrete units. As illustrated in Figure 4.3, these divisions could each represent one of eight different values. An encoding device imposes the discrete set of values of, say, signal amplitude. These values then represent the coded signal as transmitted. The receiving device transforms these discrete units into some interpretable form. This could be binary signals, i.e., a series of discrete 0-1 signals, or the discrete units could be used as a code (or formula) to reproduce an original analog wave.

Among the advantages of pulse code modulation is that it allows for an interface between usual analog forms of communications and the binary code of digital computers. Also, as you may have learned from the marketing of

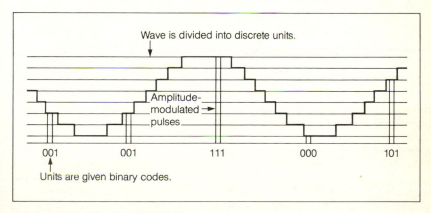

Figure 4.3 *Pulse Code Modulation*

compact disks, digital forms of coding can be used to obtain relatively distortion-free recording or transmission.

The trend is to create telecommunications systems—telephone, television, or music recording—that employ digitizing of signals. This is a reflection of the increasing convergence between the processes of computing and telecommunicating.

Binary Codes

Essentially, a *binary code* is an ensemble of 0 and 1 (two-state) signals used in digital communications. In computer communications the ensembles, called words, are of equal length in a given computer system, and each variation in their constituent elements (i.e., the pattern of 0 and 1) reflects the coding of some symbol such as a number, letter, or special referent. (Each such word, made up of *bits,* constitutes a *byte* of information.) Suppose, for example, that the system used 8-bit words (i.e., each byte contains eight bits); then the system for coding decimal numbers is as follows:

Binary	Decimal
00000000	0
00000001	1
00000010	2
00000011	3
00000100	4
.	.
.	.
.	.
01110000	112

There is more to be said about binary codes when we discuss computers (Chapter 5).

INTERACTIVE CAPABILITIES

Two-Way Transmission

As we noted in Chapter 1, interactive communication implies that two-way message traffic can travel on the same communications channel. A telephone conversation is a common example of true interactive telecommunications. By contrast, a commercial radio broadcast is strictly a one-way channel. Although this contrast is simple enough, there are further variations, such as citizen's band radio, in which only one party can be transmitting at a time (thus saying "over" to invite a response). There are also various forms of *store-and-forward* communication systems, such as electronic mail, in which communication exchanges may eventually be interactive, but at different points in time.

Duplex Transmission

When a communication link allows messages to go two ways at the same time, it is said to be *duplex,* or *full duplex.* A typical telephone connection is full duplex. Both parties can speak at once, although there may be no great advantage to this. *Half-duplex* refers to channels in which messages can go only one way at a time. Many types of data communication channels are half-duplex because there is no need for simultaneous two-way transmission.

You have experienced the feel of a half-duplex system if you have used equipment whereby individuals in a single setting are communicating by telephone conferencing to an individual in another location. If you pick up a microphone to make a comment, the caller is cut off while you talk. In practice, you learn to give cues as to when you are finishing a statement so as to invite the other party to make a contribution (as with CB radio). The same effect is felt in real-time computer communications with another individual. You type your comments, then typically enter a few line spaces to indicate it is the other person's turn to communicate.

Store-and-Forward Interaction

Forms of interaction are possible even if two parties cannot immediately communicate back and forth to one another. Exchanges of ordinary postal mail with another person are certainly a form of interaction, although the time lag between messages may be days if not weeks. In telecommunications, however, the time lag in such forms of interaction can be cut down to minutes, or better yet, set for the convenience of the participants.

In store-and-forward computer conferencing, or electronic mail, your message is stored in a central computer until called for by the recipient. You might find yourself interacting with another individual with delays ranging from only a few minutes to many days, depending upon your mutual desires. However, in many computer communications systems, if you and the other party are both on-line exchanging messages, you may have the option to switch to real-time (or half-duplex) interaction.

Computer teleconferences are typically a store-and-forward interaction in which individuals communicate their remarks on prearranged topics and then other participants log-in to read those remarks and contribute any of their own. In many such systems, it is possible for individuals to send private messages to one or a selected number of participants. Teleconferences of this type are often very convenient in that they can substitute for face-to-face meetings (particularly those of a mainly informational nature). You can participate when you are best prepared for it, perhaps in the evening or early morning hours. Moreover, you can join from any point where you can connect a computer to a telephone, for example, while traveling.

SWITCHING

Circuit Switching

As the term implies, *switching* refers to the ability to connect different communications sources with destinations and to change these connections easily. The telephone network is switched; cable television is not.

Switching is necessary in any major point-to-point telecommunications system. As illustrated in Figure 4.4, imagine the consequences if all users of a telecommunications network had permanent circuits to one another, as was the case in the very early days of the telephone. With only a few users, this is no problem. For example, two users need only one circuit and four users only six circuits. But five users would need ten circuits and ten users 45. Soon the number of required circuits becomes astronomical. The solution, as also illustrated in Figure 4.4, is to connect each user to a central office *(switch)* where circuit connections can be made at a caller's request. This is the essential use of switching: to create an ad hoc pathway for a message to travel in a telecommunications system.

For many years, telephone switches were manually operated as operators literally connected calling parties to one another. (This still occurs in simple hand-operated office switchboards: a PBX or private branch exchange). Later, switching became automated as electromechanical devices, controlled by dialing codes, made the connections. The most modern digital switches are entirely computer controlled. They are far faster and more reliable than electromechanical ones.

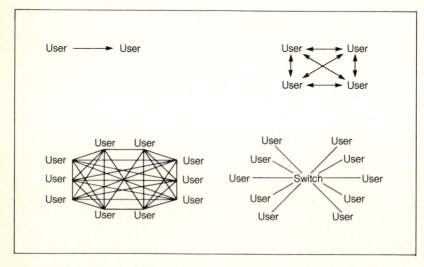

Figure 4.4 *Simple Networks and Switching*

Packet Switching

The most modern type of digital switching, called **packet switching,** does not connect circuits at all. Messages are sent in segments (packets) each of which has routing instructions incorporated within it. In rough analogy to the way a letter is routed in a traditional postal system, messages are routed by computers at circuit junctions according to their "addresses." Unlike traditional switching, there is no physical changing of circuit connectors. Packet-switching represents one of the important examples of convergence between computing and telecommunications technologies.

FROM NETWORKS TO ENVIRONMENTS

While examining the details of telecommunications, including the growth of new technologies, it is important not to lose sight of the overall implications for human communication. One of these, introduced at the outset of this volume, is that our telecommunications links, or networks, are becoming ubiquitous, putting us on the threshold of new opportunities in communication. This is the formation of a vast telecommunications environment.

The evolution of telecommunications began as people made simple extensions of their natural capabilities for communication by sound and sight. The importance was that our predecessors were able to extend their communication links by applications of technology—the science of the practical. (Surely we must include transportation as a part of this history, for until the telegraph, most communication was extended beyond our natural abilities only by the transportation system.)

Even by the early years of the twentieth century, given the remarkable growth of telegraphy, the coming of the telephone, and radio, our human capability for making communications links had increased immeasurably. In the middle years of our century, television, sophisticated telephony, data networks, microwave relay systems, electronic computers, and the coming of the communications satellite and broadband cable systems have marked another large increase in the number of our communications alternatives. Now, as we approach the twenty-first century, these systems, often themselves complex networks, have the potential for interlinkage through digital technologies to form one vast omnipresent communications environment. That we are entering an era when it is possible to access the powers of modern, technology-based computing and telecommunications systems by tapping into this network is surely a significant milestone in human history. Indeed, the network is more than a telecommunications system; it is our new *environment*.

Topics for Research or Discussion

━━━━━━━ Conduct a brief investigation of the alternative long-distance telephone services in your area. What are the price differences that distinguish them? Do most of these services compete for the same customers or are some more attractive to businesses and others to households? What do you think the future holds for these competing companies?

━━━━━━━ Most telecommunications experts would agree that fiber optics are a "cutting-edge" technology in modern communication networks. However, in the practical world of telecommunications, there is often controversy about the benefits of fiber optics over coaxial cable technology. Investigate this comparison to reveal the contrasts between the two. You may wish to write your report as if it were a consulting assignment for a company or institution that must make a decision between the two technologies.

━━━━━━━ A *modem* (modulator-demodulator) converts digital computer messages into analog signals for transmission on the telephone network. Prepare a brief report on how this is accomplished.

━━━━━━━ The ancients can be credited with the development of certain telecommunications applications—for example, smoke signals, semaphores, drum beating, and fire beacons. Examine the history of communication for these examples and others. What were the major functional uses? What types of codes were used? What are some of the major functional contrasts with modern telecommunications?

━━━━━━━ You may recall from this chapter that multiplexing involves sending numerous messages simultaneously over the same telecommunications network. Conduct a brief research project into modern methods for multiplexing. What are the current advances? What has allowed these advances? What new applications are now feasible (e.g., sending voice and data messages simultaneously over the same telephone line)?

References and Further Readings

Bargellini, P.L. "Commercial U.S. Satellites." *IEEE Spectrum* (October 1979): 30–37.

Bowers, R. "Communications for a Mobile Society." In *Telecommunications Policy Handbook,* edited by J. Schement, F. Gutierrez, and M. Sirbu, Jr. New York: Praeger, 1982, 275–306.

Bowers, R., A.M. Lee, and C. Hershey. *Communications for a Mobile Society: An Assessment of New Technology.* Beverly Hills, Calif.: Sage, 1978.

Business Week. "Telecommunications" (October 11, 1982):60–66.

Dordick, H. *Understanding Modern Telecommunications.* New York: McGraw-Hill, 1986.

Dorros, I. "Evolving Capabilities of the Public Switched Telecommunication Network." In *Telecommunications Policy Handbook,* edited by J. Schement, F. Gutierrez, and M. Sirbu, Jr. New York: Praeger, 1982, 11–26.

Fombrun, C. and W.G. Astley. "The Telecommunications Community: An Institutional Overview." *Journal of Communication* 32 (1982):56–68.

Hiltz, S.R. and E. Kerr, eds. *Studies of Computer-Mediated Communications Systems: A Synthesis of the Findings. Final Report to the Information Science and Technology Division, National Science Foundation (IST-8018077).* East Orange, N.J.: Upsala College, 1981.

Hiltz, S.R. and M. Turoff. *The Network Nation: Human Communication Via Computer.* Reading, Mass.: Addison-Wesley, 1978.

Martin, J. *Future Developments in Telecommunications.* Englewood Cliffs, N.J.: Prentice-Hall, 1977.

Martin, J. *Telecommunications and the Computer.* 2d ed. Englewood Cliffs, N.J.: Prentice-Hall, 1976.

Meadow, C. and A. Tedesco, eds. *Telecommunications for Management.* New York: McGraw-Hill, 1984.

Quaglione, G. "The Intelsat 5 Generation." *IEEE Spectrum* (October 1979):38–41.

Singleton, Loy A. *Telecommunications in the Information Age: A Nontechnical Primer on the New Technologies.* Cambridge, Mass.: Ballinger, 1983.

5

Computing

Among the greatest advances of our age has been the ability to construct information processing machines. Essentially, a computer is an information technology that can carry out a set of stored instructions (program) for the analysis and manipulation of data. In this chapter we examine the fundamental aspects of electronic computing.

TOPICAL OUTLINE

Computers as Intelligent Circuitry

Fundamental Components
Central Processing Unit
Memory
Input/Output

Codes, Programs, and Languages
The ASCII Code
Stored Instructions
Major Programming Languages
Systems and Applications Programs

Distinctions Among Computers
Mainframes, Minicomputers, and Microcomputers
Computer Generations

Computers and Communication

COMPUTERS AS INTELLIGENT CIRCUITRY

The inner workings of a modern electronic computer are essentially an array of integrated circuits. These are the well-known chips that compress, on a tiny piece of pure silicon, all the microscopic circuits and components, such as transistors and diodes. The design of each integrated circuit has a particular function as a component of the larger system that comprises the computer. This is what is meant by having the chips arrayed on a "board."

The majority of integrated circuits in computers have the capability to store, transform, transfer, and otherwise calculate binary information. Again, you can think of such information as being represented by the numbers *0* and *1*. In physical terms, you can visualize the 1 versus 0 signal of this electronic code as a distinction between voltage levels, as shown in Figure 5.1.

Although binary codes are typically used to represent complex information, we can, nevertheless, consider simple communication systems that are limited to a two-state code. For example, differentiation between the red and green signals at a railroad switch is a meaningful message to the train engineer as to whether or not to proceed. (Some of the early history of telecommunications involved such applications to railway systems.)

To accommodate more than a single two-message differentiation, computers combine data circuits; so at any one time there can be an array of the values 0 or 1. For example, a system with two data lines could accommodate four different messages, or eight messages with three data lines and sixteen with four lines. As you may already know or have noticed from this example, the capacity of binary code systems increases by the power of two for each line that is added.

Many home computers of the late 1970s and early 1980s used eight data lines to form an internal communications system or *bus* to link the main components of their system. (That's why they call them *eight-bit systems*.) Seven of these lines were the basis for a binary number ensemble (we shall discuss the role of the eighth shortly) representing numbers, letters, special characters, and computer control codes. These meaningful ensembles are typically referred to as a **byte,** a fundamental unit of information. (Byte also refers to several other concepts.) Perhaps you recall that the popular Apple II+ computer was advertised as having a memory that could accommodate 64,000 bytes. Practically speaking, that meant that its memory circuits could store that many individual numbers, letters, or control codes. (In reality, the number is somewhat less because parts of this memory were used for other purposes.)

Of course, it is necessary to have a capability for controlling the flow of the on-off signals on the data bus. This is done by an additional circuit that carries a clock signal. This clock is a timer that also varies between 0 and 1. When at 1, it allows the data lines to exchange information among various components of the computer. (You also may have noticed the clock speed mentioned in computer advertisements. The faster this clock can control signals, the faster a computer can operate.) An eight-bit computer can handle 256 different signals, and they are moved among the memory, the central processing unit, and any input or output devices by the data bus.

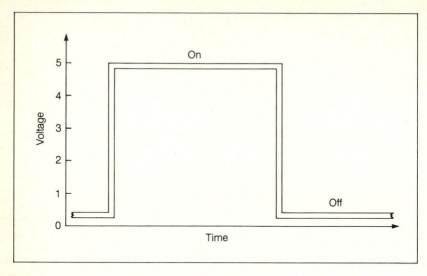

Figure 5.1 *Binary Code as a Voltage Differentiation*

By moving information according to the instructions of its operating system and whatever applications instructions (i.e., program) are loaded, a computer moves information from keyboard input, disk, tape, or memory through the central processing unit to perform operations and back to output devices (monitor screen, printer, or disk) to present or record the results. Again, all of these components are linked to the bus, which is controlled by the central processing unit.

When we design and control computer circuits to carry out our instructions (e.g., calculation, communication, text processing, simulation, gaming, and data storage and retrieval), we are, in a general sense, getting a machine to undertake designated "intelligent" operations on information. In this same broad sense a computer is a bundle of "intelligent" circuitry. Let us now look closer at how these circuits serve the certain basic functions found in any computer.

FUNDAMENTAL COMPONENTS

Central Processing Unit

As illustrated in Figure 5.2, there are three basic components: the central processing unit, the memory, and the input/out system. The central processing unit **(CPU)** is the most fundamental of the three components that make up a basic computer system. A CPU contains registers (storage units), logic units

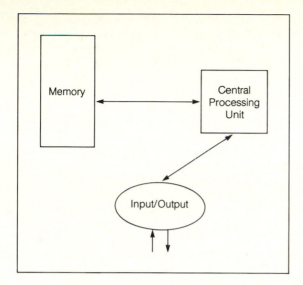

Figure 5.2 *Basic Components of a Computer*

(performing calculations), a control unit, and in-out circuits (**ports**). Altogether, the CPU takes the program instructions from memory and carries out the prescribed operations. In so doing, it temporarily receives, stores, and retrieves information, carries out the necessary decoding and encoding, and provides output. As mentioned earlier, it also controls and synchronizes the flow of information among the different components of the computer system.

Computers do not perform calculations in the same way that we do mentally. For example, when computers multiply numbers, the logic unit actually adds them very, very rapidly, so fast that the time is measured in **nanoseconds,** or billionths of a second.

For many years, CPU functions were carried out by a large variety of circuits that were expensive to construct. However, the near miracles of miniaturization have made it possible to reduce the CPU circuitry for some computers to single (or several closely integrated) chips. CPU chips are called microprocessors and are a distinguishing feature of microcomputers, or more popularly, home or personal computers.

Memory

Computer memories store basic operating systems, program instructions for particular applications (e.g., accounting), and data that are to be processed. A distinction is made between **read only** memory (**ROM**) and **random access** memory (**RAM**) in that the former includes instructions permanently stored in the computer and the latter can have instructions or data entered, modified, or erased by the user or program instructions.

When you turn on a personal computer, its ability to perform self-checks, to read a programming language, or to instruct you to load such a language can result from programs in ROM. The applications program that you enter into a computer (e.g., for word processing) and any data that are generated (e.g., your own text) are stored in RAM. It is the programming that makes a computer so flexible that it can be used for jobs ranging from numerical calculations to text processing. It is important that computer memories be able to store large amounts of information, and that such information be accessed and stored rapidly and accurately. Consequently, memory capacity and access speed are a major part of what you pay for in a more sophisticated computer system.

Data or programs can be entered into memory directly from the keyboard or from disk or tape storage devices. They can also be entered from telecommunications links from another computer.

In a very general sense, RAM and ROM are distinguished from storage devices such as disks or tapes which are auxiliary units (whether inside or outside the computer case) and typically have much greater capacity. You may have stored a variety of applications programs (word processing, accounting, and games) on disks and then selected and transferred one of them electronically from the disk to the random access memory of your computer.

Memory circuitry that is connected by the data bus with the other main central units of a computer can be likened to the pigeon holes of an old-fashioned desk. Each location in memory has a specific address at which the computer (following program instructions) stores specific information. Each location in memory holds one byte of information (comparable to one character).

So long as a computer is turned on, whatever is located at a particular memory location stays there until instructions are given to remove it or replace it. This means that you can put particular reference numbers in memory (something like memory locations on a desk calculator), and you can recall the values from those locations again and again without the need to replace the original number. For example, you might have a program instruction that says

A = 56

The computer will store A in a particular location; then whenever it is addressed, the value 56 will be read.

A special capability of the computer bus is an address function that allows access to specific memory locations. In very simple terms, you could think of this bus as a system of addressing circuits, in which any pair of wires from the two sets addresses a specific memory location. (For example, if you are the owner of a 512K Apple MacIntosh computer, the address bus is capable of accessing roughly 512,000 different memory locations.)

Input/Output

Input/output devices are probably the most visible components to a computer user in that they are the means by which people and machines exchange information. These devices include a variety of options. For example, input devices include

- Keyboards or terminals, a mouse, joystick, or trackball
- Magnetic tapes, ranging in size from cassettes to large reel-to-reel systems, floppy disks of varying sizes
- Fixed disk systems (metal disks in enclosures)
- Punched cards and punched card readers, including the keypunch machine to put coding on cards
- Optical scanning systems that can read text or other types of codes directly, magnetic character readers (often used on bank checks)
- A light pen, which writes on a light-sensitive computer screen to feed information into the computer
- Fingertip systems whereby a small wire grid beneath the screen senses touch and communicates it to the machine
- Speech recognition devices
- Graphics pads
- Telecommunications links

The most common output devices are a video screen and a printer. But there are others, for example:

- Plotters that can execute graphs or charts of different types
- Disk drives for recording processed data or text
- Telecommunications devices that transmit processed information
- Various sound-generating devices, including synthesized speech
- Punched tape

CODES, PROGRAMS, AND LANGUAGES

The ASCII Code

Because we use computers for more than numerical problems, it is necessary to have a scheme for representing other characters, such as letters, symbols, or special control characters. The most widespread scheme in use for such representations is called the *American Standard Code for Information Exchange* (ASCII, or "askee" as spoken). In the ASCII code, numbers, the alphabet, and other characters are represented by a seven-bit binary system. Table 5.1 illustrates some of the assignments for the ASCII code.

The ASCII code contains an eighth bit that is used as a special character for error checking. As described in computer circuitry, each ensemble of eight-bit code, since it represents a character, would be called a byte. Computers manipulate a fixed number of bits in each processing instruction. Some of the

Table 5.1 *The American Standard Code for Information Exchange*

Decimal	Binary	Letter
65	01000001	A
66	01000010	B
67	01000011	C
68	01000100	D
69	01000101	E

earlier and simpler home computers used word lengths of eight bits, whereas improved ones use a 16-bit word. (This basic number of bits used in a computer instruction is called a *word*. The longer the word the more complex the instructions a computer can process.) Most mini and mainframe computers use a 32-bit word.

Stored Instructions

The early prototype computers of the 1940s were designed to perform only specific calculations. Other uses required rewiring the computer. It was John von Neumann who developed the idea of storing a set of instructions in a computer's memory in such a way that they could be easily changed for various applications. This was the *stored program computer* associated with the first generation of machine design.

The specific electronic instructions for the central processing unit of computers are sufficiently complex and removed from our natural language that they are difficult to use for direct *(machine-language)* programming, although this is done in many applications. This problem has been alleviated by the development of intermediary languages that serve as translators between the natural language of the programmer and the machine language of the computer. The computer serves as an interpreter so that the programmer can write instructions in a language that is easier to manipulate than machine code. Also the language is sometimes designed to fit the nature of the application (engineering, business, or databases). The best-known computer language, FORTRAN, incorporates this concept in its name: *formula translator*.

Major Programming Languages

There are many programming languages, most of which were developed to take better advantage of computer capabilities, to improve upon an existing language, or to better fit a particular type of application. A summary of the better-known languages is given in Table 5.2, along with notes on their typical areas of application.

One general quality of programming languages is the degree to which they approximate the natural language of users—that is, the degree to which commands resemble language or specifically, in our case, English. A pro-

Table 5.2 *Major Programming Languages*

FORTRAN, 1954
John Backus, IBM

Formula Translator. Used mostly for scientific work, with some applications to business. Used widely in the behavioral sciences for numerical analysis.

LISP, 1958
John McCarthy, MIT

List Processor. For artificial-intelligence applications. Based on making connections between lists and tables of data.

ALGOL, 1959
Peter Naur and committee

Algorithmic Language. Good for scientific uses. Designed over a period of years by international group of computer scientists.

COBOL, 1959
Committee

Commerical and Business-Oriented Language. Used mostly for business applications. The goal was to develop a standard language.

JOVIAL, 1960
Jules Schwartz, System Development Corp.

Used almost exclusively for military systems, for example, in air defense computer software.

APL, 1962
Kenneth Iverson, Harvard, IBM

A Programming Language. Good for modeling programs. First used by IBM.

PL/1, 1964
Bruce Rosenblatt, George Radin, Standard Oil, IBM

Programming Language-1. Designed for the IBM 360. Used mostly in business applications.

BASIC, 1965
John Kemeny and Tom Kurtz, Dartmouth

Beginner's All-purpose Symbolic Instruction Code. Result of a Dartmouth decision to provide time-shared computing available to all students, which led to a need for an easy-to-use programming language.

PASCAL, 1971
Niklaus Wirth, Eidgenossiche Technische Hochschule, Switzerland

Named for Blaise Pascal, inventor of an early calculating machine. Used in many modern applications. Advocates have felt that it would replace BASIC for microcomputers, but to date it has not.

---

gramming language that closely approximates a natural language is called a *high-level* language. An interesting contrast is that the higher the level of the programming language (easy for you to use), the more complex is the computer's task in the translation. Lower-level languages leave much of the translation work up to the programmer and make far fewer demands on the computer. (Perhaps you have a popular computer program that is written in machine code and therefore takes up much less space in your computer's memory and runs much faster than a program coded in BASIC, which would require the computer to interpret the program prior to its operation.)

Although most computer languages are quantitatively oriented, various nonquantitative applications have given rise to alternative programming strategies. For example, LISP (for *list processing*) is a qualitative programming language that has been very important in artificial intelligence research. There are also high-level languages that are used for database manipulation, in which little calculation is required. For example, in dBase III, a popular database program for microcomputers, the command to retrieve certain information from a data file is a simple, Englishlike:

DISPLAY last-name, phone-number

Although some programming languages may *seem* nonquantitative, all eventually reduce to a quantitative type of operation at present. One of the anticipated advances in computers (see later discussion on the fifth generation) is a machine that has a nonquantitative architecture.

Of course, most computer users are never faced with direct manipulation of a programming language. As they often say in the computer business, you do not have to know how to program a computer to use one. Most major applications, from mainframes to personal computers, use programs that have required extensive investment for development and testing and have been designed to make their use as easy as possible for the computer operator. In fact, the term *transparent* is applied to software as well as other engineering applications. It means that a program should be so efficient that it hardly interferes between you and what you want your computer to perform.

Systems and Applications Programs

A distinction is often made between *systems* programs and *applications* programs. Systems programs are used for the basic housekeeping operations of a computer such as memory addressing, the flow of information to and from storage devices, and keyboard communications. A particular systems program (or operating system), in turn, accommodates only its own family of applications programs. Applications programs are the programs that you may install for accounting, word processing, database management, calculations of various kinds, or telecommunications.

One barrier to wide-scale use of computers has been that each of the various manufacturers has adopted a different operating system for its computers. Thus a program developed for one system will not run on another unless expensive adaptations are made. This hampered the early growth of microcomputers. When the IBM personal computer began to dominate the market, its operating system became a de facto standard. The current trend toward deregulation in the communications fields has in part caused the lack of standards in computer hardware and software.

Programs for computers are often referred to as **software,** whereas the physical electronic structures of computers and peripheral devices are known as **hardware.** However, when programs are installed in nonerasable read-only memory chips, they are sometimes referred to as **firmware.**

DISTINCTIONS AMONG COMPUTERS

Mainframes, Minicomputers, and Microcomputers

The history of electronic computers is mostly that of mainframes. Early mainframes cost millions of dollars and were mostly for experimental purposes. For example, the Mark I, designed by Howard Aiken, an American engineer at Harvard, required the space of a gymnasium. This computer was a combination of electrical and mechanical parts. A well-known example of the first all-electronic computer was built at the University of Pennsylvania in 1946 by John Mauchly and J. Presper Eckert. This computer was called the *electronic numerical integrator and calculator,* or *ENIAC*. It weighed over 30 tons and required some 18,000 vacuum tubes and a special air-conditioning unit. It was, however, 300 times faster than the Mark I. Some have estimated that, as opposed to hand calculations, it could solve a problem in two hours that might ordinarily take 100 engineers, each working eight hours a day, an entire year to solve.

Computers like the ENIAC are referred to as first-generation computers, in that they pioneered the use of electronics as the basis for computing. All such large computers have traditionally been called mainframes. These mainframes cost in the hundreds of thousands to millions of dollars and are mostly used in government, the military, large businesses, universities, and banks. The telephone company is also a large user of mainframe computers for telecommunications routing (switching). Mainframe computers are often linked to a network of terminals so that users can communicate with the mainframe not only at the computer center but also from adjacent offices or other parts of the nation or world via the telecommunications network.

Because mainframes have such high capacity and speed, they can do many different jobs simultaneously. If you are connected to a mainframe via a terminal, you might feel as if you are the only person interacting with the machine because the mainframe can accommodate so many users at once **(timesharing).**

Minicomputers are smaller than mainframes, and their costs range from around ten thousand to several hundred thousand dollars. A minicomputer does most of the jobs of a mainframe, although it cannot handle such large amounts of information nor accommodate as many simultaneous terminals. Most minicomputers are found in businesses, banks, and large organizations such as universities.

Microcomputers are known popularly as personal or home computers. Their price is usually less than $5000, although in the price wars of the 1980s, some were obtainable for less than $100. Most microcomputers can accommodate only one user at a time, although the more expensive and newer models are approaching the capabilities of minicomputers in accommodating multiple terminals. Microcomputers are particularly useful in businesses for handling small, individualized applications, such as word processing, evaluating financial data, maintaining mailing lists, or accounting.

The distinguishing feature of the microcomputer is that its central processing components are typically concentrated on a single chip (**microprocessor**). Usually the distinction between a home computer and a personal computer is that the former is more useful for game and educational programs—it displays in color, probably has a variety of sound capabilities, can be affixed to an ordinary television set, and has a wide variety of educational and game-type programs. This type of computer tends to dominate in school use. The business microcomputer, by contrast, usually has its own monitor, an 80-column screen, relatively high-capacity dual disk drives, and a wide variety of business-oriented programs available.

One trend in the use of home and personal computers is to link them via a modem to access other computers, data bases, and information services such as news, electronic mail, and airline schedules.

Some analysts see the lines blurring among micros, minis, and mainframes in that minis are becoming sufficiently powerful to take on many of the capabilities of older mainframes. So, too, have desktop computers increased in power, so that the newest generation of high-performance desktop computers actually have the performance capabilities of a minicomputer. The lines are blurring also in another sense: computers of all sizes are becoming increasingly linked so that a person working at a micro can occasionally connect to mini or mainframe to attack more complex problems or to transfer data from a larger file down to a micro for specialized analysis.

Computer Generations

The first four generations of computers can be classified as follows:

1. Electronic vacuum tube computers
2. Transistorized computers
3. Integrated circuit computers
4. Very large scale integrated (VLSI) computers

More details on these generations are given in Table 5.3.

Researchers are concentrating on the VLSI generation of computers in the

Table 5.3 *Computer Generations*

First Generation (1946–60)

Hardware: Vacuum tubes
Software medium: Switching wires, magnetic drum or mag core memory, punched card
Software implementation: Manual or media input of machine codes (strings of numbers)

Second Generation (1958–67)

Hardware: Transistors
Software medium: Mag tape, hard disks
Software implementation: Procedural languages. More English-like commands that were translated by the computer into the number strings it could understand.

Third Generation (1964–77)

Hardware: Integrated circuits
Software medium: Floppy disks, bubble memory
Software implementation: Nonprocedural languages. Even more English-like. Commands are often in guise of answers to questions computer asks; commands translated into other commands several times before being translated into the numbers computers can understand.

Fourth Generation (1971–   )

Hardware: Microchips
Software medium: Software-on-a-chip
Software implementation: Extensible and English-like language. Capabilities of language change with the functions invoked. Computer automatically checks instructions for reasonableness. Untrained people can operate computers without learning complex vocabulary.

---

last years of the 1980s. The general design of all four generations is based on the von Neumann machine, named after the computer planner and mathematician, John von Neumann. It is composed of central processor (program controller), memory, logic unit, and input/output devices. Von Neumann machines operate largely in serial fashion, step by step. This is the aspect in which the fifth generation will show the greatest change: It will abandon the serial processing technology. There will be new parallel processing architecture, new memory organizations, new programming languages, and new operations wired in for handling symbols as well as numbers. The fifth generation will stand apart, not only because of its technology but also because it is conceptually and functionally different from the first four generations.

The new machines may be known as *knowledge information processing systems (KIPS)*. This term signals a shift from mere data processing to a highly "intelligent" processing of knowledge. These new machines are to be designed specifically for artificial intelligence functions: symbolic manipulation and reasoning. Since almost all of the thinking in the world is done by reasoning, not calculating, there is a need for computers that match this type of information processing.

COMPUTERS AND COMMUNICATION

As we have seen, information moves within a computer in a parallel fashion along the circuits of a bus, much like a trunk line. In some cases, when signals are moved outside the computer, they are still carried in parallel fashion, such as to a printer or other peripheral device. But for the most part, computer signals are rearranged into a serial sequence for purposes of data communication.

How then are these signals that travel in parallel accommodated on a single telephone line? What is done is to convert the parallel ensemble into a sequence or series of bits that are transmitted one at a time with a division between bytes. This is accomplished by special circuitry (a *communications card* on a personal computer) that connects into a computer's bus and in turn disseminates a serial sequence. One main standard for computer serial communication is called the RS-232 interface. This standard allocates the particular types of signals needed for serial communication to the different pins of a communications cable. What is necessary next is to transform this sequence of serially communicated bits into pulses for telephone line transmission. The equipment that accomplishes this transformation is called a **modem,** short for modulator-demodulator.

Topics for Research or Discussion

━━━━━ It is often pointed out by industry analysts that there are fewer microprocessors used in computers than in other devices and machines. Conduct a brief research project into the uses of microprocessors in automobiles. Consider not only such new applications as computerized maps, but how microprocessors, for example, are used in fuel injection systems. After completing your analysis, try to classify in general or abstract terms the types of functions (such as command, control information, storage, or calculation) that are reflected in the applications.

━━━━━ In the latter part of the 1980s, powerful personal computer workstations were announced by several major companies and by universities that had participated in developing them. The aim of such workstations is to provide the student not only with ample desktop computing capabilities, but also with the ability to network with other workstations and major computing facilities on campus. Examine current literature on this topic and prepare a report on the essential characteristics of these workstations. Consider information on several workstations, then compare them as examples. (TIP: The publications of EDUCOM will be useful for this project; check with your library.)

▬▬▬ The control of robots by intelligent microelectronic technology represents an example of computer manipulation of objects in three-dimensional space. In a number of educational applications, relatively simple (and sometimes interesting looking) robots are used to demonstrate the basic requirements for robotic operations—for example, navigation. Examine contemporary literature on robotics, particularly that which spells out the essential requirements for control and sensing. Prepare a brief report on the types of microelectronic applications that serve these ends. (TIP: Educational or popular computing literature may offer you some of the most interesting literature on this topic.)

▬▬▬ It is often said that the transistor, invented at Bell Laboratories, launched the microelectronics revolution. Conduct a brief research project on the development of the transistor. Essentially how does the transistor differ from the vacuum tube it replaced? What were some of the applications envisaged for the transistor? What is the distinction between the transistor and modern integrated circuits?

▬▬▬ LISP is a computer language that is nonquantitative; it is used in expert system and artificial intelligence applications. Prepare a brief report on the nature of this programming language, giving examples of certain applications. Why is it an important language for the aforementioned applications?

References and Further Readings

Anacker, W. "Computing at 4 Degrees Kelvin." *IEEE Spectrum* (May 1979): 26–37.

Bernhard, R. "Small Business Machines." *IEEE Spectrum* (January 1982).

Bernstein, J. *The Analytical Engine: Computers—Past, Present and Future.* New York: William Morrow and Co., 1981.

Blundell, G. "Personal Computers in the Eighties." *Byte* 8 (1982): 171–182.

Bulthuls, K., M.C. Carasso, J.P.J. Meenskert, and P. Zalm. "Ten Billion Bits on a Disc." *IEEE Spectrum* (August 1979): 26–33.

Business Week. "Artificial Intelligence: The Second Computer Age Begins" (March 8, 1982): 66–75.

Dertouzos, M. and J. Moses, eds. *The Computer Age: A Twenty-Year View*. MIT Bicentennial Studies, Vol. 6. Cambridge, Mass.: MIT Press, 1980.

Duda, R. and E. Shortliffe. "Expert Systems Research." *Science* 220 (1983): 261–268.

Englebart, D. *Augmenting Human Intellect: A Conceptual Framework. Summary Report*. Menlo Park, Calif.: Stanford Research Institute, 1962.

Feigenbaum, E.A. and P. McCorduck. *The Fifth Generation: Artificial Intelligence and Japan's Computer Challenge to the World*. New York: Signet, 1983.

Kerr, E. and S.R. Hiltz. *Computer-Mediated Communication Systems*. New York: Academic Press, 1982.

Leduc, N. "Communicating through Computers." *Telecommunications Policy* (September 1979): 235–244.

Nau, D. "Expert Computer Systems." *Computer* 16 (1983): 63–85.

Peat, F.D. *Artificial Intelligence: How Machines Think*. New York: Simon & Schuster, 1985.

Rochester, J.B. and J. Gantz. *The Naked Computer*. New York: William Morrow and Co., 1983.

Weizenbaum, J. *Computer Power and Human Reason: From Judgment to Calculation*. San Francisco: Freeman, 1976.

Wessel, M. *Freedom's Edge: The Computer Threat to Society*. Reading, Mass.: Addison-Wesley, 1976.

III

APPLICATIONS CONTEXTS

The greatest insights into the human aspects of the new media are found in psychological or sociological studies of use. In the next sequence of chapters we examine selected current issues regarding behaviors ranging from the use of personal media such as the telephone to the use of broad applications such as public media and national or regional development. In all cases, the fundamental concern is for understanding more about how technology can extend or otherwise enhance the human capability for communications. In addition to reviews of these issues, each chapter in this section contains first-hand reports written by researchers who specialize in studies of new media.

Personal and Social Communication

Some media technologies directly extend the range of person-to-person communication. The nature of personalness in such exchanges is the initial topic of this chapter. This is followed by a discussion of the telephone as technology for interpersonal communication, including an example of research into how individuals "value" their different uses of this instrument.

TOPICAL OUTLINE

ON THE CONCEPT OF *Interpersonal*

Impersonal versus Personal Communication

Modern theorists (e.g., Miller, 1976, 1978) often consider *interpersonal* communication as something more than simply person-to-person or point-to-point communication. If two-person communication is especially personal—i.e., if directed toward individual needs, interests, or personal motives rather than role-related motives—it is considered as a fundamental basis for the development of an interpersonal relationship, a basic social unit.

In the interpersonal process, friendships may develop with the sharing of personal information leading to feelings of mutual trust and interdependence. In slightly more detail, we can contrast personal and impersonal communication in terms of variables of motives, personal perception, information exchange, and the cumulative patterns of the exchanges. Figure 6.1 summarizes this process.

By contrast, in an impersonal exchange the motives for communication are almost exclusively associated with the object of the transaction, such as concluding a purchase, gaining requested information ("One cheeseburger please!"), making a bank deposit, or securing some service ("Pass the salt."). Although there could be degrees of personalness involved—such as individual needs in a situation—the object of communication is the transaction, not the person.

However, for a variety of reasons we may wish to "get to know" another person. That person may be important to us because cooperation is necessary so that we may achieve a goal. The other person may be the type of individual we wish were our friend or may be sexually attractive to us. In all such cases, the motive for interaction wholly or partly involves developing a relationship with the other individual. Even if another type of transaction is involved, such as a remark about the weather while mingling at a cocktail party, the motive may be to legitimize the verbal exchange. Indeed, a social gathering or a date may be organized around activities that sanction small talk that sets the stage for personal exchanges.

Among the important technological concerns related to interpersonal communication are how media enable us to extend our personal communication behaviors and their importance in the use of communication for social ends. The telephone, for example, offers us interpersonal gratification.

The Concept of Social Presence

A frequent inquiry about point-to-point technologies is whether they depersonalize message exchanges. Most point-to-point media are narrower channels, so to speak, than the full range of audio, visual, tactile, and even olfactory stimuli of physically proximate exchanges. For example, the telephone denies us the exchange of symbols of the nonverbal, visual code. If we

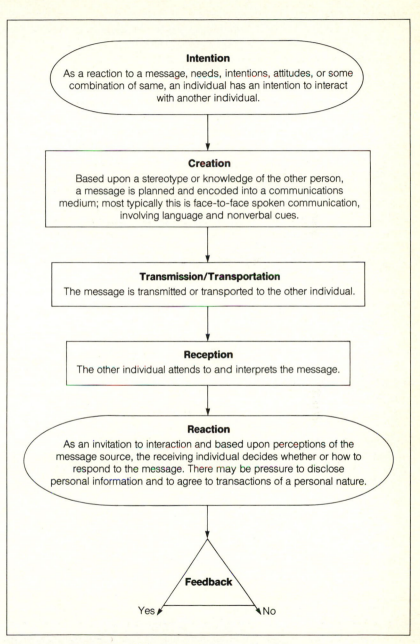

Figure 6.1 *Interpersonal Communication as a Social Process*

cannot see how a person is reacting to our conversation, it may be more difficult for the interpersonal communication process to develop. Yet we can make a telephone conversation very personal. We compensate for the lack of visual emotional cues by using our vocal (paralinguistic) ones. Even more so, we can compensate for the restrictions of the medium by using language that is especially personal.

Psychologist John Short and his colleagues (Short, Williams, and Christie, 1976) have studied how different media affect what they call "social presence." This concept is equated with feeling that a communications experience is sociable, sensitive, warm, or personal. To the degree that a medium restricts nonverbal cues and is not interactive and private, it lessens one's feelings of social presence. For example, Short, et al. found that people in a business setting rated written correspondence lower in social presence than telephone conversations, and both considerably lower than face-to-face conversations.

Again, however, we can compensate for the restrictions of a medium by making our language choice or message structure more personal. In fact, this is often a part of the training given when individuals are taught to use technologies such as teleconferencing, telemarketing, or computer mail. They are given advice on how to compensate for the restrictions of the medium.

Personal Gratifications

Personalness in point-to-point communication can also be examined in terms of media usage—that is, in the functions that users claim for a particular message exchange. This was a clear finding in Charles Steinfield's study (see his report) of the use of electronic mail in a large corporation. Although the system had been installed presumably for business uses, a much larger component than expected of mainly social messages evolved. It seems that if we have a desire for personal exchanges, we can adapt the medium to our message. Similar findings—although anecdotally revealed—seem to hold for customers' uses of the major on-line computer communications services (e.g., CompuServe and The Source) and local electronic bulletin boards. There are large amounts of informal social exchange (an observation not unfamiliar to persons experienced with amateur radio or the CB radio explosion of a decade ago).

Point-to-Point Telecommunications

Interpersonal communication typically assumes a one-to-one context as, for example, a conversation. This communication is interactive in the sense that both individuals are simultaneously involved in creating and responding to communication. Although the natural occurrence of interpersonal communication is in a face-to-face situation, there is a long tradition of using media to extend personal human contact (e.g., personal correspondence). Usually we call this point-to-point communication as contrasted with person-to-person communication. Examples of point-to-point communication include

- traditional mail
- electronic mail

- telephone
- telegraph
- paging
- CB radio
- facsimile

THE TELEPHONE AS A SOCIAL INSTRUMENT

A Direct Extension of Personal Communication

Probably no communication technology is a more direct extension of interpersonal communication than the telephone. "Reach out and touch someone" is more than an advertising slogan; it describes an important communication gratification. Some 96 percent of all households in the United States have telephones, a fact that makes this technology as nearly ubiquitous as radio and television, which have similar penetration. By the same token, there are vast differences in the availability of the telephone in different cultures in the world. For example, on the continent of Africa, the percentage of penetration is approximately one-tenth of one percent. Even the communist countries of the world, although industrialized, lag behind their Western counterparts with a penetration of about 20 percent.

Paradoxically, despite its near-century of usage in the world, the telephone has seldom been researched for its social uses as a communications medium. There are studies of telephone installations, traffic patterns, and costs, but very little has been researched about how we use the telephone in our daily lives. Recently the contribution of the telephone to social and economic development has been viewed as an especially important topic of study. Outlined in an essay by Dordick, Phillips, and Lum (1983) are four priority areas for research on the social uses of the telephone:

- There is a need to make more informed decisions about telephone investments policies for countries, especially countries with developing economies.

- The telephone has a distinct role in the social reorganization of a society. That is, electronic networks allow flexibility in the reconfiguration of what were formally geographic or spatial constraints upon organizations or groups. If we are able to reconfigure our social lives, businesses, or even governments, without such constraints, what are the consequences?

- As we enter an era of new media in the home—especially cable television, personal computers, and satellite dishes—there are many opportunities for combinations or competition with the telephone network. What services will be offered and how will people choose among them?

- It is important that we learn more about the values that individual users place on the availability of telephone services. As described in the research example in the next major section of this chapter, there may be various dimensions of gratification underlying a person's use of the telephone. These gratifications may place different demands upon the network and could be a basis for assigning costs to services.

Universal Service

Although consumers use the phone mostly for social purposes, they also place a high value on it as an emergency or business-transaction instrument. In the current period of divestiture, the long time cross-subsidy in which business-sponsored, long-distance service typically subsidized inexpensive home service is coming to an end. This means that there may be cutbacks in the traditional concept of universal service. A question emerges as to which customers or what types of services should be cut back, and by whose criteria? Should the poor and the elderly be given a subsidized, *life-line* service, and if so, who will pay for it? Should emergency service be universally available and inexpensive, as contrasted with possible rate increases for extensive use of the telephone for everyday social conversations?

The current assumption that a free enterprise telecommunications market is best for a democratic society requires close examination. The problem is that with deregulation, it is not clear who must pay the bill for subsidized services.

Potentially, issues such as these should be of great interest to the public. However, judging by American public's apathy toward telecommunications matters, it is possible that they will simply accept changes in their service rather than be influential in shaping its future.

One reason for public ignorance is that much of the news about telephone deregulation has been interpreted only in the business pages of the daily press, and not from the social or the home consumer standpoint. Inadequate coverage of issues crucial to the home telephone user, especially in terms of the availability of different services or the rights of the consumer, go un-highlighted. The fact that many telephone consumers do not now understand how their bills are calculated, or the confusion over alternative long-distance services, is evidence of the problem of public ignorance. One very realistic outcome of deregulation is that the telephone will be much less of an instrument available for universal service. Although equity in social issues will continue to be respected, it is likely that the final shape of America's telephone system may be most of all affected by economic realities.

GRATIFICATIONS FROM TELEPHONE USE: A RESEARCH EXAMPLE

Attitudes about Telephone Use

There may be no profound reason why the telephone has seldom been the object of social-psychological research other than that most communication researchers have been more interested in print, broadcasting, and film. The research to be described here is one of a series of studies that eventually led to a questionnaire for investigating the detailed attitudes that consumers hold toward the telephone and, in particular, for assessing correlates of these attitudes.

Our research (Williams and Dordick, 1985) in this area began with relatively homogeneous groups of undergraduate students. Of course, students are hardly representative of any large population, yet they were interesting to us for several reasons. First, we knew from discussion groups that the students were substantial users of both local and newly offered long-distance telephone services and were opinionated about both. Consequently, if any group should yield a set of identifiable attitudes toward telephone usage, we would expect to find them with students. Second, several of our clients in the telephone business were particularly interested in university students because they represented a new generation of valued customers.

A major assumption underlying this research was that, with the advent of deregulation in the United States, consumer attitudes toward the telephone were becoming increasingly complex. Presumably, this is one consequence of having new types of billing, the need to make decisions about telephone equipment, and the choice of long-distance services. There was also consideration in the first year of deregulation that the popular press had carried considerable coverage of consumer reactions to telephone service and rates.

We assumed that the time had passed when the telephone was thought of as a simple utility and paid for monthly in a manner similar to electricity, water, or natural gas. Accordingly, we expected that telephones, telephone service, and perhaps even the telephone industry would be increasingly relevant "purchases" to consumers. Moreover, as their personal choices were exercised and their monthly bills paid, customer attitudes toward telephone concepts should become more salient.

A Research Strategy

The basic approach in this program of research consisted of the following:

1. To use focus group techniques to ferret out concepts and attitudes relative to the telephone's use.
2. To develop a questionnaire for gathering and quantifying data on these concepts.
3. To conduct selective analyses to assess the likely dimensions of concepts and attitudes underlying the detailed questions.

Eventually a 150-item questionnaire was developed. The intent was to gather as much information as possible on all aspects of the students' thoughts and attitudes about their telephone use and then later to reduce this data by statistical procedures to fewer concepts and dimensions. The questionnaire had the following main sections:

I. Demographics
II. Descriptions of their parents' homes
III. Equipment and services now used
IV. Specific attitudes about telephones and services
V. General attitudes about telephone service
VI. Knowledge of selected services
VII. Likely uses of the telephone

Dimensions of Usage

Although our interest in the initial analysis was based on analyzing responses to individual questions, a broader, more theoretical aspect of this analysis will be described here. The analyses were focused on two questions.

1. Will correlations among responses to questions about telephone use suggest more fundamental dimensions of usage?
2. If such dimensions of usage can be identified, to what degree are they predictable from other components of the questionnaire data, for example, demographics?

Responses to questions regarding usage from 325 students were subjected to factor analysis. This was an analysis which in practical terms was an attempt to see what broad, relatively independent factors might be subsumed in the data. Six factors were extracted in the analysis, accounting for 52 percent of the variance. Four of these factors were quite easily interpreted as corresponding to social, emergency, and transactional uses and to a more general use akin to conducting business. Examples of questions under each of these factors were as follows.

■ *Social:* To stay in touch with a friend; to relieve boredom; to communicate with someone out of state; to tell someone you love them or miss them.
■ *Transactional:* To make a reservation; to purchase a ticket; to make an appointment.
■ *Emergency:* To summon help.
■ *Business:* To ask parents for a loan; to maintain a business relationship.

For the next step in the analyses, questionnaire data were combined for the three dimensions of social, transactional, and emergency usage. (Because of the small amount of variance accounted for, business was collapsed into the transactional factor.) Multiple regression analyses indicated that these factors could be predicted from other information obtained from the respondents. Results were as follows:

■ *Social use:* Approximately 20 percent of the social use factor score was predictable from general attitudinal questions that related to the importance of the telephone in the student's life. Further predictor variables included available spending money and living off campus.

■ *Transactional use:* Again, the general attitude of the importance of telephones on various questions contributed to the prediction of the importance of transactional uses. This was followed by several questions regarding equipment and usage, such as having an extension phone, sharing the phone with others, and positive attitudes about phone service. Again, roughly 20 percent of the variance of transactional use factor scores was predicted from various answers to questions.

■ *Emergency use:* General attitude questions that phones are important and, in this case, that service is satisfactory were the best predictors of rating of the need for emergency services. Only 11 percent of the variance in emergency use was predictable from combinations of other questions.

Our conclusion was that the general attitudes about telephone services may indeed reflect more fundamental dimensions or perceptions of types of service needs. Given the further examination of this model or variations of it with further populations, we felt that researchers could devise a strategy for comparing different consumer populations, age groups, and consumers living under different consumer conditions. The research strategy could also be used as a basis for comparing service attitudes in different countries. In all, we felt that the research added to our ability to conceptualize attitudes about telephone usage and service.

From a slightly more technological standpoint, the three-dimensional model drew attention to the point that various needs such as social, transactional, and emergency uses of the phone impose quite different demands upon the network. For example, very little time may ever be required for placing emergency calls; the important quality is that the customer has the service immediately available at any time of the day or night. On the other hand, social uses of the phone might involve extended periods of usage, and there may be more tolerance for encountering a busy signal or having to attempt to contact the other party several times before a connection is made. Somewhere between is the transactional type of call, which one may make much more frequently than an emergency call but would probably be calling during business hours and would want to be able to make an immediate connection.

Another question was whether importance to the consumer and cost to the network could be equated. There may be a basis in this equation for considering a differential pricing scheme based upon value to the consumer. A final point of interest was that when (or if) data services become important to home consumers, what types of usage will be important to them, and will these types be similar to the present results of the study? All such questions reflect examples of issues arising when a technology is applied to interpersonal communication.

REPORT

Personal Communications via Electronic Mail

Charles Steinfield
Michigan State University

The computer-mediated communication system is often viewed as the cornerstone of the office of the future. A growing body of literature supports the general finding that managers, professionals, and other information workers spend the majority of their work day in person-to-person, communication-related activities. Electronic mail systems can support this communication and, because of the unique features of this new medium, can help overcome many of the inefficiencies of face-to-face, telephone, and traditional mail or memo-based interaction. Furthermore, this new office technology may do more than simply make communication more efficient. Many qualitative changes in the ways individuals communicate at work can result from the introduction of an electronic mail system.

Just how do individuals in organizations make use of electronic mail, and what factors affect use of the system? A study undertaken within a large information processing firm investigated these issues, providing many insights about the likely uses of computer-mediated communication in large organizations.

Several hundred users of the company's electronic mail system took part in this study. First, participants described their use of electronic mail by noting the degree to which they employed the system for a variety of purposes. Contrary to much previous research on electronic mail, a broad array of usage purposes were observed, including both task-related and socio-emotional uses. Most respondents reported major use of the system for task-related information exchange. Electronic mail appeared to be an effective way to distribute company memoranda, to seek information from others in the organization, and to give and receive feedback on reports or ideas.

A number of attributes make this medium appropriate for exchanging information, particularly when the information is complex. For example, users compose messages in writing and have time to think about what they wish to say without distraction by responses from recipients. In addition, information is stored for future display. Respondents, in fact, reported that keeping a record of interactions was a relatively common purpose of use. This permanence, however, may have imposed restrictions, as very limited use of the system for confidential communication was evident.

Other frequent task-oriented uses included scheduling meetings and appointments, coordinating project activities, and monitoring project progress. These further illustrated how an electronic mail system could facilitate the performance of organizational tasks.

But not all communication at work is task-related, nor is the use of electronic mail. The system in the study was used for such social purposes as keeping in touch with contacts around the company, participating in entertaining conversations, and playing games. Many people belonged to groups or were on distribution lists that reflected shared interests or hobbies. It was not uncommon for employees to turn to these electronic clubs during lunch or breaks for stimulating interaction or even light reading. This also helped people keep track of what was happening around the company.

The electronic mail system enabled certain forms of interaction that might not have been possible, or at least practical, by other media. One fairly common use represents a new form of information seeking. Respondents used distribution lists to send (broadcast) requests for information to large numbers of employees on the system. Similar but less frequent applications, which also took advantage of the ability to broadcast messages, were the use of the system to conduct opinion polls and to advertise or respond to products for sale.

One example of broadcasting demonstrated the potential power of an organizational electronic mail system for access to remote sources of information. A design team in one city facing a critical decision sent a message throughout the company documenting the possible design choices. Within hours, they had received several hundred responses, with an overwhelming majority favoring one design option over another. This team then fed back the survey responses to users and opted for the majority choice. All of this took place in one work day. No other communication technology could have allowed such rapid access to this organization's remote technical expertise.

Once these patterns of use were established, the research turned to identifying factors that seemed to affect task or social use. It appeared that the amount of task use was related to three sets of factors. First, an appropriate electronic mail infrastructure must be created. Greater task use was positively related to the number of respondents' coworkers accessible via the system. Easy access to terminals for connecting to the system was critical. The more that people were forced to share terminals the less their task use. Second, a clear communication need must exist between users. Geographic dispersion of coworkers normally inhibits task communication and creates a need to use the system. Persons with coworkers in other locations were greater task users. Finally, major users had a positive orientation to electronic mail and, in particular, perceived it to be an effective means of communication.

Social use, on the other hand, was best predicted by demographic characteristics such as organizational tenure. In particular, the newer and younger employees were more likely to employ electronic mail for socioemotional purposes. A positive orientation toward the system was also important. Rating of the perceived sociability or social presence afforded by electronic mail correlated highly with social use.

It appeared that new employees used the system to help become integrated into the informal social networks in the organization. (It was noted that the establishment and maintenance of interpersonal relationships could have payoffs to the organization even though interaction was initially social.) Social networks could facilitate the flow of task-oriented information, as new employees learn where to turn when faced with task-related questions. Indirect benefits resulted from the increased cohesiveness and morale that was fostered by network interaction.

Although generalizations beyond the organization under study were difficult to make, this study did demonstrate the versatility of computer-based communication in a setting where few restrictions are placed upon use. As an integral component of office automation, electronic mail not only enables more efficient and easier access to human information resources, but from this study it appears that it can help make the office of the future a more personal workplace as well.

Further Readings

Hiltz, S.R. and M. Turoff. *The Network Nation: Human Communication Via Computer*. Reading, Mass.: Addison-Wesley, 1978.

Hiltz, S.R. *Online Communities: A Case Study of the Office of the Future*. Norwood, N.J.: Ablex, 1984.

Kerr, E. and S.R. Hiltz. *Computer Mediated Communication Systems*. New York: Academic Press, 1982.

Rice, R. "The Impacts of Computer-Mediated Organizational and Interpersonal Communication." In *The Annual Review of Information Science and Technology,* edited by M. Williams. Vol. 15. 1980.

Uhlig, R., D. Farber, and J. Bair. *The Office of the Future: Communications and Computers*. New York: North Holland, 1979.

Topics for Research or Discussion

████████ Construct a brief questionnaire on social presence qualities perceived in a telephone conversation. An example of an item on this questionnaire could be as follows:

SOCIABILITY: high _____: _____: _____: _____: _____ low

Also include such scales as sensitivity, warmth, and personalness. Make arrangements through a mutual friend so that six or so individuals have this questionnaire, then rate your social presence when you call them. What generalizations can you draw about your telephone image? What do you think the reasons are for this image? How might you improve it?

▬▬▬▬▬▬ Some say that communications technologies dehumanize the way we communicate. Computer-based types of communication are usually regarded as examples of this. What is your own analysis of this situation? If, indeed, some of the more human aspects of communication are filtered out of the message, what types of compensations might be implemented to overcome the problem?

▬▬▬▬▬▬ Relative to other communications media, there is relatively little social or behavioral research on the use of the telephone. Conduct a brief library investigation to see what titles you can find in this area. Examine several of the volumes and note the areas of concern or research. From what you have found, as well as your own thoughts on the matter, what type of social research do you think should be conducted in this area? (TIP: Start with the books by Ithiel de Sola Pool listed in the references to this chapter.)

▬▬▬▬▬▬ In the 1960s, citizen's band (CB) radio became so popular that the Federal Communications Commission could not keep up with the request for licenses. Examine the literature about CB radio, including how it began, and attempt by examination of newspapers and magazines of the 1960s and 1970s to describe the CB craze. Conclude your paper with speculation on why CB radio ceased to be so popular.

▬▬▬▬▬▬ As described briefly in this chapter and elsewhere in detail by the author (Frederick Williams and Herbert S. Dordick, *Social Research and the Telephone* [Los Angeles: Annenberg School of Communications, 1985]), values that customers place on telephone service may depend on their particular functional usages. The authors say that functional usage may divide into social interaction, business transactions, and emergency use. Conduct a brief survey of six or so of your classmates, inquiring into their uses of the telephone over the past week. Given these uses, inquire as to the relative importance they ascribe to these uses. What generalizations can you draw about how they value telephone service?

References and Further Readings

Card, S., T. Moran, and A. Newell. *The Psychology of Human-Computer Interaction*. New York: Erlbaum, 1982.

Dizard, W. "Reinventing the Telephone." *InterMedia* 11(1983):9–11.

Dordick, H., P. Lum, and A.F. Phillips. "The Social Uses of the Telephone." *InterMedia* 11(1983).

Miller, G.R. "The Current Status of Theory and Research in Interpersonal Communication." *Human Communication Research* 14(1978):164–178.

Miller, G.R. *Explorations in Interpersonal Communication.* Beverly Hills, Calif.: Sage, 1976.

Muson, H. "Getting the Phone's Number." *Psychology Today* (April 1982):42–49.

Panko, R. "Electronic Mail." In *Advances in Office Automation,* edited by K. Tackle-Quinn. New York: John Wiley, 1984.

Pool, I. de Sola. *Forecasting the Telephone: A Retrospective Technology Assessment.* Norwood, N.J.: Ablex, 1982.

Pool, I. de Sola, ed. *The Social Impacts of the Telephone.* Cambridge, Mass.: MIT Press, 1977.

Reid, A. "Comparing Telephone with Face-to-Face Contact." In *The Social Impact of the Telephone,* edited by I. de Sola Pool. Cambridge, Mass.: MIT Press, 1977, 386–415.

Short, J., E. Williams, and B. Christie. *The Social Psychology of Telecommunications.* New York: John Wiley, 1976.

Steinfield, C. *Communicating Via Electronic Mail: Patterns and Predictors of Use in Organizations.* Unpublished Ph.D. dissertation, Univsersity of Southern California, Annenberg School of Communications, 1983.

7

Group Communication and Teleconferencing

We do not communicate in groups the same way we do as individuals. This is particularly true when participants are linked by media technologies as in the case of *teleconferencing*. This chapter describes generalizations developed over the last decade about such conferencing, including computer linkages.

TOPICAL OUTLINE

Communication in Groups

Types of Teleconference Configurations

Some Generalizations Regarding Teleconferencing
Major Observations
Gratifications from Teleconferencing
Participation Styles
Predicting Use

Trends in Teleconferencing

COMMUNICATION IN GROUPS

Typically, we consider groups in terms of face-to-face contexts that include speech and nonverbal communication as the main media. The small group process, long a topic for social scientific research, is summarized in Figure 7.1.

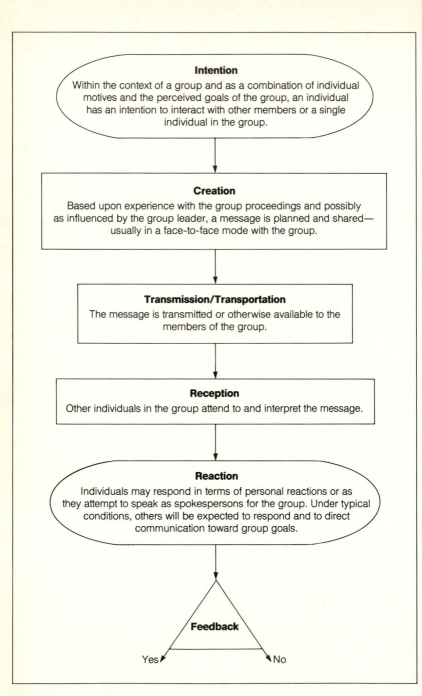

Figure 7.1 *Small Group Communication*

Research has given us many insights into small group communication, including the effects of different styles of leadership, group pressures, interactions of group purposes, and patterns of participation. However, in all such applications, the participants are physically proximate. The communication is face-to-face. Our question is how these group processes change when individuals are linked by various communications media. What are the impacts on the processes described in Figure 7.1?

TYPES OF TELECONFERENCE CONFIGURATIONS

Teleconferencing applies to almost any communication system that people use to exchange information according to some agenda so as to reach a goal. For many years, teleconferencing has been possible in the audio mode by connecting more than two parties in a telephone conversation. For over a decade audio teleconferences could be combined with video links, either full motion or slow scan. Recently, teleconferencing has included the process by which individuals can confer either by store-and-forward or by direct message exchanges of computer text. Table 7.1 summarizes the major configurations for teleconferences.

SOME GENERALIZATIONS REGARDING TELECONFERENCING

Major Observations

Although teleconferences can represent the many configurations described above and vary widely in purposes, participants, or topics, a few generalizations have emerged from research and practice, including the following points.

1. Savings in cost and travel time are given as the main corporate reason for implementing or sponsoring teleconferencing.
2. Teleconferences require special planning. For example, individuals who know one another personally are usually more successful than participants who begin as strangers. If a conference is to be among strangers, time should be spent getting acquainted.
3. Teleconferencing does not work equally well for all purposes. For example, it is usually more useful for achieving informational or managerial goals than for bargaining or negotiation.
4. When other factors are favorable, teleconferences can be more efficient than face-to-face meetings because there is less opportunity for small talk and digressions.
5. Positions of power sometimes change in the teleconferencing environment because of some person's adeptness in using the medium or in the organiza-

Table 7.1 *Types of Teleconference Configurations*

Audio

This is the traditional multiparty telephone line for voice communications. It can link as many individuals as the group wishes but may become cumbersome if you expect over 15 people to interact.

Enhanced Audio

Audio can be combined with a type of visual information exchange, such as facsimile, computer exchange, electronic mail, telewriting, data transmission, electronic blackboard, and remote slide control.

Video: Full Motion

This can be one- or two-way video (and audio) using the full bandwidth as does commercial television. It is expensive, and broadband transmission capability is required.

Video: Slow Scan

Bandwidth demands for full-motion video can be reduced to regular telephone lines for transmitting still pictures at about one frame or picture every 30 seconds. This is usually augmented with an audio channel.

Computer Teleconferencing

Messages can be easily exchanged via interactive computer-to-computer communication. Although this can be used for real-time exchanges (i.e., direct interaction), many examples of computer teleconferencing involve store-and-forward systems. They also include software for assisting in the format, organization, and indexing of the conference.

---

tion of the conference. Some individuals gain power simply because they appear more favorably on camera (the "Hollywood effect").

Gratifications from Teleconferencing

Video conferencing, when studied in a corporation, has been found to be effective for situation-specific communication (Dutton, Fulk, Steinfield, 1982). But its study otherwise becomes problematic when compared with other technologies because it is not at the stage where it can be used by anyone as part of everyday activities. Computer teleconferencing, electronic bulletin boards, and special communication networks such as The Source, on the other hand, are becoming more accessible as more people purchase home computers.

   The choice of a conferencing system depends on many factors, including the degree to which the need is task related, the desire to establish and maintain contacts outside one's own geographic area, or the need for access to data bases. Research (Hiltz and Turoff, 1979) has indicated that there is an affective dimension to both "real time" and "store-and-forward" computer teleconferencing. The observation is that people anthropomorphize computers and use them for everything from psychiatrists and confessors to partners in crime or priests. Hiltz (1978) and Phillips (1983) found that people are very active in their socioemotional uses of computers for working out problems and using face-saving tactics or stream of consciousness thinking, even when conferences are task oriented. It can be inferred from these and other studies emerging in the literature that there are many different types of gratifications associated with computer-mediated communication that have not yet been assessed.

Participation Styles

How does a teleconferencing medium affect the emotional dimensions of participant style? This was the topic of Phillips' (1983) study of the transcripts of three computer conferences, the contents of which were analyzed with emphasis on whether (and if so, how) the medium's special characteristics seemed to enhance or diminish the emotional dimension of participation. Results indicated that the ability to communicate while alone at a terminal, in a written mode, and without the expectation of immediate feedback, encourages stream of consciousness communication of thoughts that exhibit spontaneity and creativity. Once users surmounted the initial period of frustrations associated with learning new procedures, there was a tendency to be very open, conversational, and personal in writing style, with numerous instances of emotional expressiveness, humor, metaphorical language, and overt sociability.

The necessity of having to write down thoughts did not seem to negatively affect the amount of participation by some users, although several individuals expressed discomfort with the literary and computer skills required for a computer conference. Other participants mentioned that lack of immediate feedback was somewhat disturbing. In general, however, the study revealed a predisposition towards acceptance and of positive attitudes toward computer mediated communication.

Predicting Use

The idea that stereotyped attitudes might be used to predict the probability of adoption of a technology has been another avenue of teleconferencing research. An example are the studies conducted in conjunction with the establishment of a teleconferencing system by the Atlantic Richfield Company (Ruchinskas and Svenning, 1981). The researchers examined individuals' intentions to use video conferencing as well as factors predicting use of current communication options in an 800-employee study spanning ten operating companies of this corporation.

Use of available communication alternatives (telephone, face-to-face, and writing) was predicted primarily by current work activities and cross locational communication needs (Ruchinskas, 1982). When considering use of a new communication option such as video conferencing, employee attitudes and benefits assume a much more prominent role (Svenning, 1982). Beliefs about teleconferencing's attributes and expected benefits appear to be particularly critical to whether or not employees intended to use this new technology (Svenning, 1983). "Cross locational" (i.e., among different offices) communication needs became more influential in explaining the frequency of projected video conferencing use. Svenning and Ruchinskas (1984) suggest that although beliefs, fears, or expectations about new communication technologies will influence initial or trial usage—that is, communication behavior—the cross locational communication requirements and work activities will be more influential in predicting routine use.

TRENDS IN TELECONFERENCING

Probably the most major trend in teleconferencing is its acceptance as a regular mode of information exchange. Many organizations now employ teleconferencing in a routine manner. At the same time, there must be visible supporting motives for teleconference implementation, because the general advantages of substituting communication for transportation are too vague to support major investments. More often, teleconferencing is implemented where it very clearly fits an organization's needs, as, for example, a Monday morning schedule briefing, launching a new product to a national network of dealers, or including the remote participation of a distinguished speaker at a conference. Other trends include

- Greater use of inexpensive media alternatives, such as telephone conferences (perhaps accompanied by mailed visual materials)
- The availability and use of new technological alternatives for visual image transmissions, such as slow scan video or facsimile, may become especially popular as costs drop
- An increase in companies offering teleconferencing equipment or facilities such as satellite dishes, camera and voice equipment, and video displays
- Rapid growth as the use of personal computers as communications terminals increases and as the availability of public data communications services increases
- In the slightly more distant future as voice-and-data terminals become available, this dual mode of teleconferencing will increase.

REPORT

Computer-Mediated Teleconferencing

Amy F. Phillips and Pamela Pease
University of Southern California

Background

Research in computer-mediated communication is unique and exciting because the technology allows both asynchronous and synchronous information exchange and the sending of public or private messages. Nontraditional methods like on-line monitoring can be combined with surveys and interviews. Researchers are also challenged to deal with such issues as the potential invasion of the privacy of an electronic community and the extent to which their own participation as observers may alter the nature of the system being studied. This case study of the XCOMM (not the real name) system was conducted to learn how computer teleconferencing is being used as an educational tool in an electronic classroom environment. Over a six-month period we gathered data from remotely located students in an advanced adult-education course on innovative approaches to upper-level management and strategic planning.

XCOMM is a private adult-education institution that currently uses the facilities of a larger computer communication network to teach seminars. Students can join on-line conferences or enter comments off-line in a text-editing mode. After an initial face-to-face orientation session, further instruction occurs via computer.

The instructors, who are experts in their fields, each teach a one-month seminar: lecturing, leading discussions, and assigning reading. The students hold high-level positions in corporations and academic institutions that are concerned with futures research and management in high-technology industries. Although everyone is encouraged to participate, no formal grading or mechanism exists for ensuring involvement. There were 26 students, four of whom were female. Their average age was 49 years. Nine students reported having previous experience using computer-mediated communication systems.

Research Overview

To study a new interactive medium in an instructional context, we had to pay attention to the way the system was used to gratify different needs. Unlike traditional media, computer teleconferencing combines task-oriented and personalized communication by individuals and groups in interpersonal transactions that may accommodate different writing styles. We wished to assess the students' attitudes toward the system as it was used for learning, personal communication, and social interaction. The research questions were the following:

1. Is the computer teleconferencing medium structured to meet successfully the educational needs of professional adults?
2. Is the medium perceived as being used for both the communication of personal thoughts and the expression of complex topic-related ideas?
3. How is the medium used for interpersonal-like social and group interaction?

Method

Multiple research methods were used to compile the most complete data possible. Baseline demographic data were gathered from a short survey given to students at the initial orientation meeting. At the beginning of the seminar an on-line questionnaire was sent to each student's electronic mail-box that contained open- and closed-ended questions regarding reasons for enrolling and attitudes toward and expectations of a computer education system.

Telephone interviews were conducted halfway into the seminar semester to obtain in-depth information about the survey responses and to provide additional insights about issues that concerned the students. Throughout the seminar, we acted as passive on-line observers of the transactions and conducted informal analyses of written transcripts to gain a broad picture of the electronic classroom environment. Future research would include more formal content analyses of the transcripts.

Results

Attitudes toward the System

Generally, the students were motivated more by the prospect of using a new form of communications medium than by the educational rewards they expected to receive. Attitudes toward the system were ascertained from the on-line questionnaires and telephone interviews. They are presented as answers to the three primary research questions.

1. Educational Effectiveness
    - The computer conferencing medium was perceived as being very useful for the diffusion of new theories and points of view and the accumulation of knowledge.
    - The least successful aspect was the lack of a formal structure to provide regular feedback and exchange of ideas between students and instructors.

- The instructors were not seen as being successful facilitators who could assure everyone's active involvement.
- There was dissatisfaction with the lecture-type teaching strategy; simulations would have been preferable for promoting "the flow of creative juices."

2. Computers as a Communications Medium

- The communication process was rated highly; it was felt to be a good medium for the expression by individuals of personal and complex topic-related ideas.
- Occasional mechanical problems inhibited use.
- The medium is not considered a "magic substitute for or improvement over other written media"; computer communication was often supplemented with other forms of contact such as telephone and mail.
- The medium can be used to improve one's personal writing style.

3. Social and Group Interaction

- The quality of interaction was often considered variable and uneven.
- Interaction was "fantastic when people are actively involved" but quickly became "old hat" when no new or controversial topics were introduced to encourage more lively discussion.
- A certain "in-group" was perceived as often dominating the interaction; others became merely passive readers.
- The system was valuable for making new friends and expanding social networks.

Issues Related to Educational Computer Conferencing

In the course of observations and interviews, important issues not directly related to the research questions were discerned. These issues, which deserve further study, are presented here.

1. Personal, emotional communication is evident and expected.

- Students expressed their hopes to form intimate friendships, to find pleasure in communicating, and to gain confidence through using this medium.
- Students frequently used the public conferencing system to express private emotions like anger, joy, humor, and frustration.

2. Personal feedback and evaluation are needed.

- The lack of a formal feedback mechanism presented a barrier to increased participation, resulting in feelings of loneliness and isolation.
- Due to a lack of direct cues, personal attention via *any* medium from instructors or other participants would have been appreciated: "If I

had only received one phone call during the seminar, one example of someone truly caring about me, I would have felt better about being more active."

3. Fear of ostracism.

 • There were complaints of feeling ostracized for expressing views that opposed those of the most verbal participants; this was conveyed by ignoring dissenting comments.

 • Some students desired a forum for "half-baked ideas," brainstorming, and controversial discussions; they were disappointed. Others preferred the medium for "protection" and made "rehearsed, edited comments, not spontaneous."

4. Status consciousness.

 • Perceptions of a high-status "in-group, an elite A-Team" were found, making it hard for new members to break in and achieve equal status.

 • Status was often accrued more by association with innovative communications technology than by real involvement in the seminar.

5. Novelty as a motivator.

 • Active participation decreased as the novelty of the new technology, concept, and content of seminars wore off.

 • New challenges, simulations, and better educational strategies that incorporate "other-worldly roles" are needed to maintain interest and personal involvement.

6. Barriers to research: the researcher as secret agent.

 • Serious objections were raised to the presence of researchers, who were feared to be invading the privacy of students.

 • The nature of teleconferencing is a crucial issue. Is a conference open due to the availability of written transcripts or private and subject to tapping like the telephone?

 • Do researchers have the right to open what may be considered a closed, private community? The participant observer may be perceived as an intruder, violating very private space, and even analogous to a computer hacker breaking into a system!

Conclusion

This exploratory study shows that computer teleconferencing has the potential to succeed as a tool for adult education. Very serious issues, both substantive and methodological, have been posed as challenges to innovative researchers. The successful adoption of the medium and generalizability to other types of user groups will depend upon continued evaluation and creative solutions.

References and Further Readings

Pease, P.S. *Long Distance Training for Maine and New Hampshire's Vocational Rehabilitation Counselors.* Madison, Wis.: Center for Interactive Programs, University of Wisconsin, 1983, 39–44.

Phillips, A.F. "Computer Conferencing: Success or Failure?" Vol. 7, *Communication Yearbook,* edited by R. Bostrom. Beverly Hills, Calif.: Sage, 1983, 837–856.

Rice, R.E. "Impacts of Organizational and Interpersonal Computer-Mediated Communication." In *Annual Review of Information Science and Technology,* edited by M. Williams. White Plains, N.Y.: Knowledge Industry Publications, 1980.

Rice, R.E. and D. Case. "Electronic Message Systems in the University: A Description of Use and Utility." *Journal of Communication* 33, (1984): 131–152.

Topics for Research or Discussion

▬▬▬▬ Teleconferences can involve a wide range of media, such as telephone, one-way video with audio feedback, two-way audio-video, facsimile transmission, or computer-based text exchange (see Table 7.1). Some of these options are more valuable for some kinds of teleconferences than others. For example, exchange of complex quantitative information is difficult under an audio-only channel. Prepare a brief paper that describes a variety of teleconferencing goals; then for each describe the ideal media configuration.

▬▬▬▬ Suppose that you were planning an audio-video teleconference among individuals who did not know one another and that you wanted a high degree of social presence. What would you do to try to achieve this goal? How would you know whether you were successful?

▬▬▬▬ Many local telephone systems can now accommodate three-party calls. Set up a mini teleconference among several of your friends. Try to accomplish some specific objective such as planning a meeting or deciding upon a weekend activity. After the conference, interview the participants about their impressions of the process. What were the main differences from a face-to-face meeting? What could improve the process?

▬▬▬▬ The work of Ruchinskas and Svenning (see references in this chapter) involved the attempt to predict the use of teleconferencing by managers and technical personnel in a large corporation. Select an example of an organization of your choice. How might you go about predicting use of teleconferencing? What configurations (Table 7.1) would be optimal? How would you motivate use?

■■■■■■ An important line of research is to examine how known generalizations about small group communication processes are affected when media separate the participants. Take one such generalization from this literature (e.g., styles of leadership) and speculate on the likely impact of a teleconferencing environment.

References and Further Readings

Baird, M. and M. Monson. "How to Tackle Training for Teleconferencing Users." *Educational and Instructional Television* 14(1982):45–50.

Carey, J. "Interaction Patterns in Audio Teleconferencing." *Telecommunications Policy* 6(1981):304–314.

Charles, J. "Approaches to Teleconferencing Justification." *TElecommunications Policy* 5(1981):296–302.

Danowski, J. "Computer-Mediated Communication: A Network Analysis Using a CBBS Conference." Vol. 6, *Communication Yearbook,* edited by M. Burgoon. Beverly Hills, Calif.: Sage, 1982, 905–924.

Dutton, W., J. Fulk, and C. Steinfield. "Utilization of Video Conferencing." *Telecommunications Policy* 6(1982):3:164–178.

Elton, M. *Teleconferencing: New Media for Business Meetings*. New York: American Management Association, Membership Publication Division, 1982.

Fowler, G. and M. Wackerbarth. "Audio Teleconferencing versus Face-to-Face Conferencing: A Synthesis of the Literature." *Western Journal of Speech Communication* 44(1980):236–252.

Gold, E. "Trends in Teleconferencing Today Indicate Increasing Corporate Use." *Communications News* (1982):48–49.

Green, D. and K. Hansell. "Teleconferencing: A New Communications Tool." *Business Communications Review* ll(1981):10–16.

Hansell, K., D. Green, and L. Erbring. "A Report on a Survey of Teleconferencing Users." *Educational and Instructional TV* 14(1980):70–76.

Hiemstra, G. "Teleconferencing, Concern for Face, and Organizational Culture." Vol. 6, *Communication Yearbook,* edited by M. Burgoon. Beverly Hills, Calif.: Sage, 1982, 874–904.

Hiltz, S.R. "Control Experiments with Computerized Conferencing: Results of a Pilot Study." *Bulletin of the American Society for Information Science* 5(1978):11–12.

Hiltz, S.R. "Experiments and Experiences with Computerized Conferencing." In R. Landau, J. Bair, and J. Siegman, eds., *Emerging Office Systems.* Norwood, N.J.: Ablex, 1982, 182–204.

Hiltz, S.R. and M. Turoff. "The Evolution of User Behavior in a Computerized Conferencing System." *Communications of the ACM* 24(1981):729–751.

Hiltz, S.R. and M. Turoff. *The Network Nation: Human Communication by a Computer.* Reading, Mass.: Addison-Wesley, 1978.

Hoecher, D. "A Behavioral Comparison of Communication Using Voice Switched and 'Continuous Presence' Videoconferencing Arrangements." Murray Hill, N.J.: Bell Laboratories, 1978.

Johansen, R. "Social Evaluations of Teleconferencing." *Telecommunications Policy* 1 (1977):395–419.

Johansen, R. *Teleconferencing and Beyond: Communications in the Office of the Future.* New York: McGraw-Hill, 1984.

Johansen, R., R. DeGrasse, Jr., and T. Wilson. "Group Communication Through Computers." Vol. 5, *Effects on Working Patterns.* Menlo Park, Calif.: Institute for the Future, 1978.

Johansen, R., K. Hansell, and D. Green. "Growth in Teleconferencing— Looking Beyond the Rhetoric of Readiness." *Telecommunications Policy* 5(1981):289–303.

Johansen, R., R. Miller, and J. Vallee. "Group Communication Through Electronic Media: Fundamental Choices and Social Effects." *Educational Technology* (August 1974):7–20.

Johansen, R., J. Vallee, and K. Spangler. *Electronic Meetings: Technical Alternatives and Social Choices.* Reading, Mass.: Addison-Wesley, 1979.

Krueger, G. and A. Chapanis. "Conferencing and Teleconferencing in Three Communication Modes as a Function of the Number of Conferees." *Ergonomics* 23(1980):103–122.

Noll, M. "Teleconferencing Communication Activities." *Proceedings of IEEE,* 1977, 8–14.

Olgren, C. and L. Parker. *Teleconferencing Technology and Applications.* Dedham, Mass.: Artech House, 1983.

Phillips, A.F. *Computer Conferences: Success or Failure?* Vol. 7, *Communication Yearbook,* edited by R. Bostrom. Beverly Hills, Calif.: Sage, 1983, 837–856.

Pye, R. and E. Williams. "Teleconferencing: Is Video Valuable or Is Audio Adequate?" *Telecommunications Policy* (June 1977): 230–241.

Remp, R. "The Efficacy of Electronic Group Meetings." *Policy Sciences* 5(1978):101–115.

Rice, R.E. "Communication Networking in Computer Conferencing Systems: A Longitudinal Study of Group Roles and System Structure." Vol. 6,

Communication Yearbook, edited by M. Burgoon. Beverly Hills, Calif.: Sage, 1982, 925–944.

Rice, R.E. "Computer Conferencing." Vol. 2 *Progress in Communication Sciences,* edited by B. Dervin and M. Voigt. Norwood, N.J.: Ablex, 1980, 215–240.

Rice, R. "The Impacts of Computer Mediated Organizational and Interpersonal Communication." In M. Williams, ed., *Annual Review of Information of Science and Technology.* White Plains, N.Y.: Knowledge Industry Publications, 1981.

Ruchinskas, J. "Communicating in Organizations: The Influence of Context, Job, Task, and Channel." Ph.D. dissertation, University of Southern California, 1982.

Ruchinskas, J. and L. Svenning. "Formative Evaluation for Designing and Implementing Organizational Communication Technologies: The Case of Video Conferencing." Paper presented at the Annual Conference of the International Communication Association, Minneapolis, May, 1981.

Strickland, L., P. Guild, J. Barefoot, and S. Patterson. "Teleconferencing and Leadership Emergence." *Human Relations* 31(1978):583–596.

Svenning, L. "Predispositions Toward a Telecommunication Innovation: The Influence of Individual, Contextual, and Innovation Factors on Attitudes, Intentions, and Projections Toward Video Conferencing." Ph.D. dissertation, University of Southern California, 1982.

Svenning, L. and J. Ruchinskas. "Organizational Teleconferencing." In R. Rice, ed., *The New Media: Uses and Impacts.* Beverly Hills, Calif.: Sage, 1984.

Wilkens, H. and C. Plenge. "Teleconference Design: A Technological Approach to Satisfaction." *Telecommunications Policy* (September 1981):216–227.

Williams, E. "Teleconferencing: Social and Psychological Factors." *Journal of Communications* 28(1978):125–131.

8

Communication in Organizations

Since the 1970s we have witnessed the steady growth of organizational communication as a topic for research. Now new questions arise as many of the communication activities within organizations are assisted by information technologies in applications referred to as "office automation," "office of the future," and other such promising labels. In this chapter, we first examine the organization as a communication context, then consider how technologies have been implemented to enhance organization operations.

TOPICAL OUTLINE

Organizations as Communication Contexts
Two Levels for Analysis
The Organization as a Communications Environment
The Organization as a Communicating Entity

Technologies in Organizational Communication

Promises of Office Technologies
Clerical and Managerial
Larger-Scale Views
Conducting Needs Assessments

Studying the Implementation of Office Technology
An Example of Implementation Research

ORGANIZATIONS AS COMMUNICATION CONTEXTS

Two Levels for Analysis

There is probably no environment more communication intensive than the modern organization. Life in this environment involves a wide mix of interpersonal and group behaviors, the ability to gather information from a variety of external sources, often the use of advertising to influence the public, and the ability to use an increasingly complex variety of information technologies. For overall analyses, it is useful to distinguish between the many communications processes within the organizational environment as against a more global conception of organizations themselves as communicating entities.

The first implementations in the 1950s and 60s of information technologies were within organizations, as in the case of accounting, word processing, or manufacturing control. In the 1970s and 80s we have seen the integration of these applications into overall management information systems (MIS). As we approach the 1990s, the trend is clearly toward expansion of the management information system beyond the organization to include, for example, suppliers, customers, and banking services.

The Organization as a Communications Environment

Figure 8.1 portrays a general conception of the internal communication processes of an organization. In general terms, "communication controls work," both in terms of managers issuing directives for action and workers responding with messages of work performed. In some organizations the product or service is itself communication.

On a more detailed level it is useful to examine clerical as distinguished from managerial communications activities. The earliest and best-known applications of information technologies were on the clerical level of most organizations. Here the attempt has been made to automate or otherwise assist in the accomplishment of routine tasks. Among the tasks are

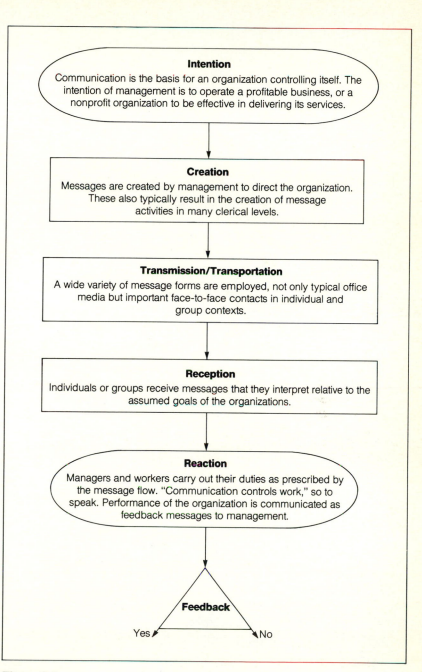

Intention

Communication is the basis for an organization controlling itself. The intention of management is to operate a profitable business, or a nonprofit organization to be effective in delivering its services.

↓

Creation

Messages are created by management to direct the organization. These also typically result in the creation of message activities in many clerical levels.

↓

Transmission/Transportation

A wide variety of message forms are employed, not only typical office media but important face-to-face contacts in individual and group contexts.

↓

Reception

Individuals or groups receive messages that they interpret relative to the assumed goals of the organizations.

↓

Reaction

Managers and workers carry out their duties as prescribed by the message flow. "Communication controls work," so to speak. Performance of the organization is communicated as feedback messages to management.

↓

Feedback

Yes   No

Figure 8.1

- Typing
- Telephone answering, routing
- Filing
- Mail handing
- Appointments scheduling
- Proofreading
- Errands
- Keeping records
- Reading
- Writing

As you can guess, many of these tasks have been assisted by office technologies. On this level, the expectation is that more of these tasks can be accomplished in less time (or for lower costs) if technologies are used. "Productivity," as applied at this level, is essentially one of increasing the ratio of output to input.

By contrast, managers are often said to operate in the environment of a *control hierarchy,* meaning that their main duties are to facilitate the work of others in the organization. The abstract concept of this hierarchy is illustrated in Figure 8.2. Some of the manager's major duties as defined on this abstract level include

- Planning
- Organizing
- Directing
- Motivating
- Monitoring
- Controlling
- Evaluating and adapting

In a more practical view, much of a manager's daily behavior involves

- Meetings
- Gathering information
- Making decisions
- Troubleshooting
- Seeking follow-up information on projects
- Drafting and reacting to correspondence
- Placing and receiving telephone calls
- Creating reports
- Acquiring or discharging employees

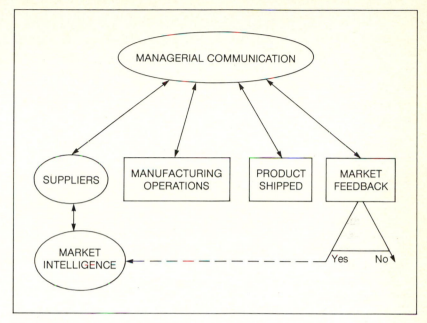

Figure 8.2 *An Abstract View of Managerial Communication*

Although it is possible to see from Figure 8.2 and the above list how technologies can facilitate the manager's work, it is less of a "work-per-unit-of-time" proposition than for the clerical worker. Herein lies one of the major problems in assessing technological impact on management: Some of the management functions are elusive. We will return to this problem later. One improvement, as outlined in most advice to new managers, is to use technology to transfer clerical tasks to clerical workers. Most managers find themselves doing clerical tasks (e.g., brief correspondence, keeping an office schedule), but with careful consideration some can be given over to the clerical force. Technology can aid in this transfer (as in setting up a form letter system so secretaries can directly respond to routine correspondence).

The greatest impact, however, of information technologies upon the manager's behavior has come from the ability to integrate existing and new technologies into management information systems. In a broad sense, this is the use of technologies to enhance the managerial tasks listed above and described more generally in Figure 8.2. As we said at the outset of this chapter, the current trend is for this "command and control" system to reach beyond the boundaries of the organization to the business or public environment in which organizations operate and in which businesses compete with one another. In fact, the ability to have control over this environment (e.g., through links with suppliers or customers) can increase the competitive advantage of one business over another.

The Organization as a Communicating Entity

Organizations cannot exist in a vacuum. Most critically depend on interaction with the environment in order to exist, especially if it is a profit-making business. Figure 8.3 illustrates some of the typical communication processes of a modern business. In these cases, we can consider the organization itself as a communicator, although individuals or groups of employees will be carrying out the actual behavior.

Information technologies have a major impact on a communicating organization because the new telecommunications networks make it much easier to gather or disseminate information from diverse sources. Also, the units of a decentralized organization can be efficiently linked by dedicated networks; in fact, this makes decentralization increasingly easy and attractive.

TECHNOLOGIES IN ORGANIZATIONAL COMMUNICATION

Even without the new communications technologies, most organizations already make extensive use of media technologies. A summary of traditional uses is compared with so-called new technologies in Table 8.1. The phrase "office of the future" is more a metaphor than an objective description of media technologies. Essentially, it means that the tools of the managers and workers are changing.

In addition to the technologies summarized in Table 8.1 are more general applications such as

- *Word processing.* Word processors or computers with word processor programs are used to replace the typewriter. Essentially, the text is kept in electronic form until all editing is performed, thus saving erasures and retyping. Also, documents that are in electronic form can be filed without having to be saved on paper.

- *Electronic mail.* When point-to-point messages are sent over electronic communications systems such as a computer network or by facsimile machines, the messages are called electronic mail.

- *Electronic funds transfer.* Computers are used to send messages regarding the transfer of funds among different accounts, to obtain balances, or to perform other desired banking operations.

- *Teleconferencing.* This is a meeting held via a telephone, audio-video, computer, or combined communications network.

- *Management information system.* This is not one piece of equipment you "buy" as such, but a configuration of existing or new office technologies in an integrated system. For example, word processors could be linked to provide electronic mail, transaction systems could generate summary data bases to be used for business analysis, or an ordering system could automatically replace inventory as well as funds transfers.

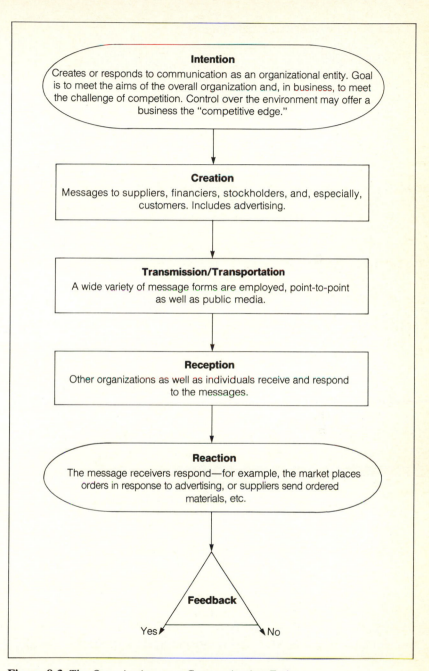

Figure 8.3 *The Organization as a Communicating Entity*

Table 8.1 *Technologies in the Office*

INPUT DEVICES

Traditional: Paper and pencil, dictation, telephone, typewriter

New: Intelligent typewriters, computer terminals, optical character reader, touch sensitive screens, "Mouse," speech

INFORMATION PROCESSING

Traditional: Manual processing, calculators

New: Facsimile machines, semi-automated processing (intelligent typewriter, intelligent telephone) and computers, PABX (Private Automatic Branch Exchange)

INFORMATION TRANSFER

Traditional: Wires, telephone network, telegraph, postal service

New: Coaxial cable, fiber optics, telephone network, microwave, satellite, broadcasting, local area network, transportation

OUTPUT DEVICES

Traditional: Typewriters, copying machines, mimeograph and ditto machines

New: Printers of various quality levels, plotters, facsimile output, video screen displays, digital readout displays, photo copiers

STORAGE DEVICES

Traditional: Paper files, microfiche, index cards

New: Magnetic tape and disk, optical disk, intelligent circuitry

---

PROMISES OF OFFICE TECHNOLOGIES

Clerical and Managerial

Most initial introductions to office technologies are in terms of their promises. For example, on the clerical level technologies supposedly

- Increase the quantity and quality of work produced
- Increase the number of transactions that can be accomplished in a given period of time
- Allow flexibility of time and space—i.e., allow flexibility as to when and where work is to be done, including the possibility of having workers work at home or in decentralized offices

Of course, one managerial benefit of technology implementation is to have more efficient clerical workers. But beyond this, we witness such promises for managers as

- The ability to supervise a larger number of employees
- The ability to gather greater amounts of data and information formerly only available via selected officers or individuals (e.g., via the data processing manager)
- Assistance for screening and interpreting information
- Assistance with decisions, including testing them
- Facilitation in creating reports, designs, and other information-intensive products
- New marketing and marketing research techniques

Larger-Scale Views

It is important that the emphasis for implementing office technologies is on creating a more effective organization than on the operation of the technology itself. That is, we are ultimately more interested in managing effectively *with* the new technologies than in simply effective management of the technology itself. In a broad view, benefits include

1. *Enhanced effectiveness of business operations*—that is, the gain of more return for what is invested.
2. *Increased competitive advantage of the business*—that is, how can it and its products be differentiated to advantage against a competitor?
3. *Enhanced value of the business itself*—that is, what is its market value?

A somewhat intangible but nevertheless very significant large-scale benefit is an increase in the alternative work styles available to clerical as well as managerial personnel. This especially includes alternatives that represent increases in quality. For example, there may be a greater potential for people to feel better about themselves and their work. By automating the most trivial and repetitious of tasks, the individual is left to concentrate on more important and often creative aspects of the work, which, after all, are what we humans contribute uniquely and probably best.

Ultimately, these benefits are gained by making managers and workers more effective. In this respect, the successful implementation of technology is far more a management challenge than a technological one.

Conducting Needs Assessments

There has also been considerable interest in forecasting uses of computing or telecommunications technologies as they may meet the needs of different employees or organizational operations. One example of this approach is the *needs assessment* formulated by Robert Johansen and his colleagues (1984). Questionnaires inquire as to the importance employees place upon different communication-related activities—for example, upon document creation and editing, intra-office mail, meetings, data storage and retrieval, or performing calculations. Respondents are also asked about the benefits they expect from technology use—benefits that may include time saving, decreased travel, more control over one's schedule, flexible work hours, or opportunity for innovativeness. The results of such questionnaires are then used to design, implement, and evaluate technology implementations.

Most such studies involve determining which important operations might be best enhanced by the application of office technologies. This question puts decisions about office technologies into a more potentially productive perspective. This is because applications will not be made just where they seem the most superficially obvious (as in word processing), but where pay-off will be the greatest. An important adjunct to the forecasting type of study is to conduct research into the process of technology implementation, the topic of the next section.

STUDYING THE IMPLEMENTATION OF OFFICE TECHNOLOGY

An Example of Implementation Research

Most experienced managers, as well as evaluators of office technologies, will agree that the process of implementing a technology is typically the most crucial factor affecting successful use. That is, *getting people to use* a new technology, not the technology itself, is the critical challenge. As a consequence, studies of the implementation process have grown in importance in this field.

In the remainder of this section, we will examine a well-regarded example of an implementation study, one reported by Bonnie Johnson, Ronald Rice, and their associates (1987) of the adoption of word processing in different office environments. In this study the research team conducted telephone interviews with nearly 200 organizations that had been using word processing technology for at least two years. In a second phase the researchers selected 60 sites for follow-up visits and intensive data gathering and analysis. The project involved information from 80 managers, 302 word processor operators, and 243 individuals who initiated text ("authors"). The general purpose of this research was to answer three questions:

1. Do organizations vary in their uses and procedures for word processing?
2. How does word processing adaptation evolve?
3. What predicts the development of adaptation?

In the broadest terms, the answers to these questions were as follows:

1. Organizations do indeed vary substantially in terms of their uses and procedures for word processing.
2. Word processing adaptations generally can be described in terms of Rogers's (1983) well-known five-stage model of the adoption process.

3. Organizational communication appears to be one of the best predictors of the degree to which word processing will be applied toward particularly beneficial ends, although there are many other factors that intervene in this process.

Four Types of Adoption

Differences in adoption among the organizations studied can be described in terms of four general categories. In examining these categories below, you'll see that the most effective adoptions involve organizational change and that *reinvention* is a critical part of the process. Essentially, reinvention as used here is the discovering of still newer ways of applying a communication technology.

1. *Word processing as typewriter*. As the label implies, this was a non-innovative use of word processing where its more sophisticated capabilities over typewriting were not realized. There was little reinvention. In such cases, there was very little managerial attention given to the implementation of word processing; employees tended to teach themselves. Nor was there a tendency for employees to share the knowledge that they did gain. In essence, there was a simple substitution of word processing equipment for typewriters, with little supervisory attention.
2. *Standard adopters systems*. The researchers used the "clockwork" analogy for this level of adoption. This was a successful adoption, but there was nothing much done of an innovative nature rather than basic use of the new equipment. There was no visible *innovation* use, nor a larger view of how further changes might enhance organizational effectiveness. The focus was on short-term efficiency. This system had supervisory attention and some sharing of information among operators. However, there was little attempt to determine methods to enhance the applications.
3. *Expanding systems*. As contrasted with standard systems, the "expanded" ones did reflect a degree of reinvention, and it often was promoted by supervisors who took a leadership role. There was an emphasis on results and effectiveness, and consideration of how the system could affect the individuals in a work unit. There was reinvention, but only at a particular organizational level.
4. *High integration systems*. Here the particular mark was a high level of reinvention—reinvention that often could be associated with the enhancement of organizational effectiveness. The key was to go beyond adoption for mere efficiency's sake—to use reinvention, rather, to improve the functional contribution of that work unit to the organization. Supervisors were important on this level of adoption, as were managers. One further observation was that the high level of reinvention appeared to result from worker interaction. Or put another way, the organization-wide communication networks were particularly beneficial for the development and dissemination of new ideas.

Reinvention

The concept of reinvention, as you have seen, was an important variable in this research. As we stated somewhat differently earlier, reinvention, generally, is an adopter's further modification and improvement in the use of an innovation following its original implementation. In the most practical terms, successful reinvention would take place if an office staff found and applied the best uses of word processing for the benefit of its organization. It is more than just the simple concept of efficiency. In the Rice and Johnson research, reinvention was measured by the number of new types of applications that evolved beyond the initial implementation. These applications could be on the managerial or clerical levels.

Reinvention turned out to be a useful concept not only in differentiating degrees of successful implementation, but also as a variable to explain these differences.

Correlates of High Integration

As mentioned above, reinvention tended to reflect the degree of organizational communication, being particularly strong, for example, where there was the "networking" of the "high integration systems." The researchers linked networking and high integration to factors that have been previously associated with the reinvention process: namely, training, encouragement of experimentation, communication, and participation in decision making. In one respect, the consequence of such communication is the increase in the capabilities of people to accomplish their tasks.

This could be compared with the similar emphasis of Strassmann (1985), who considers the key factor of successful implementation in the office to be the enhancement of the individual's contribution, whether manager or clerk. In the broadest sense, Johnson and Rice concluded that quality of organizational communication, as it is related to and presumably contributes to reinvention, is a key factor in the adoption of new word processing technology. In a larger view, this same process could be presumably generalized to office automation itself. An important correlate of this conclusion, agreed upon by many others who write and research on the topic of office automation (Strassmann, 1985; Keen, 1981), is that if information technology is to be successful, it will typically involve changes in work forms and ultimately in the organizational structures that support them.

THE CHANGING INFORMATION ENVIRONMENT

Information as a Commodity

As described by Daniel Bell in *The Coming of Post-Industrial Society* (1976), one of the hallmarks of our time is that the business of acquiring, repackaging, or disseminating information is a major organizational growth area. Businesses involved in the manufacture and sale of information technologies, or companies involved in providing information services, show a high growth in the rate of return on equity. They are attractive investments because of their earning power. Often, comparisons are made between these companies and traditional industries of heavy manufacturing, or extractive industries such as mining or agriculture. Bell's forecasts reflect mostly the growth in research and development applications, although similar arguments have been made citing publishers, computer and software manufacturers, broadcast networks, and market research firms.

Although the "information" businesses have had their failures in the last decade (e.g., home computer and video-game manufacturers), and heavy industries have had many examples of resurgence (e.g., automobile manufacturers), there remains an overall trend in the growth of information as a commodity itself. And certainly the decline of selected "smokestack" industries seems irreversible (e.g., steel production). But even more noteworthy is that great investment in information technologies is not just in the industries that go by that name, but also in the reorganization of many traditional ones. A prime example is the purchase of Electronic Data Systems (a billion-dollar information business) by General Motors, coupled with the latter's claim of developing entirely new approaches to automobile manufacturing (e.g., the Saturn project). In essence, the greatest trading in information as a commodity is in the refitting of traditional industry, as well as applications in government, the military, and, it can be hoped, education.

The Rise of "Information Work"

An obvious corollary of the growth of information industries is the increased need for the accomplishment of activities that require the acquisition, processing, and dissemination of information. There has been a general metaphor of this as "information" or "knowledge" work, a reference that received much attention in the 1970s from not only Bell's original forecasts, but also from Marc Porat's *The Information Economy: Definition and Measurement* (1977). However, this metaphor has often been confused with activities that involve the clerical handling of information, as compared with the use of information to make valuable decisions. Secretaries and file clerks are typical examples of clerical level information workers. Their contribution is the direct processing of information. Their productivity can easily be measured on a "per unit" basis.

On the other hand, a person who makes decisions with information—e.g., an executive, a researcher, or an analyst—does not so much process information as extract some type of value from it. In such cases, the assessment of productivity is more elusive: One decision might eventually be far more valuable than another one, or a value might not be visible for many years. In addition, any evaluation would require a consensus as to criteria.

But despite the problems of definition, "information" work as a general metaphor is definitely on the increase in modern organizations, and it is an opportune environment for the study of communication technologies. Our challenge is to understand how to make the best of information technologies, or as better put by Paul Strassmann (1985), to gain *information pay-off*.

One key generalization, it seems, is that in the vast majority of successful implementations, organizational change has taken place. That is to say, technology does not do the job alone; success requires making people and their organizations more effective.

Information as a "Value-Added" Activity

The more recent argument has been made (see, esp., Strassmann, 1985) that the major growth of information activities in organizations is not so much related to information as a commodity as it is information as a "value-added" activity. That is, improved use of information contributes to the effectiveness of most modern organizations, whether or not they are in the "information business." Moreover, the collective monetary share of such applications far exceeds those of the information businesses in modern national economies. A coming change in organizations, therefore, is not so much that more resources will go into information businesses, as it is that more resources will go into the "refitting" of existing industries so as to take advantage of improved use of information (and, broadly, "communication") in management. Strassmann calls this "added value" and offers it as a rationale for evaluating the contribution of information activities to the effectiveness of an organization.*

If the real information revolution is in the refitting of existing industries (including service organizations), a corollary is that competition among industries may increasingly narrow to their ability to use information to their best advantage. Although this may be a fresh concept in terms of the applications of information technologies, it is simply a rephrasing of the long-respected business adage that "management will make the difference." Indeed, management is the use of information—or more broadly, communications—to operate the organization, and the criterion is effectiveness (profit and service).

---

*A similar point is a key theme in my book with Herbert Dordick, *The Executive's Guide to Information Technology* (1983), where we advise the manager to consider communication as a "cost center" (rather than simply "overhead") to be evaluated relative to its contribution to return on overall investment.

REPORT

Why Organizations Adopt New Office Technologies

Suzanne Iacono and Rob Kling
University of California

One of the major changes in the character of office life over the past 100 years is an increasing dependence upon office technologies. New "modern" technologies have constantly been developed and implemented. Over time, they become woven into the procedures of organizations and business practices so that they appear indispensable to their users. What once was novel becomes ordinary and taken for granted.

Yesterday's office machines, such as mechanical erasers and billing machines, seem archaic. In 20 years, today's advanced office systems, which feature integrated calendars, text processing, and electronic mail accessible through multiple windows on a single-font, 12-inch (half-page) video display terminal (VDT) will also seem archaic. The term *Modern technologies* does not denote particular technologies, only temporary pointers in a continuous cavalcade of technological devices.

Messages about the promise of new technologies dominate much of the public discourse about the nature of office automation. The mass media, consultants, and vendors of computer products tend to emphasize the best promises of new technologies. Vendors play a role in creating high expectations to satisfy managerial demands for increased office productivity and decreased office costs. Even when evidence about these promises is inconclusive and contradictory, they can provide managers with rationalizations for the implementation of state-of-the-art equipment. Below, we analyze three motivating factors in the implementation of new office technologies: increased office productivity, cost-savings through staff reductions, and managers' fascination with new technologies.

Office Productivity

A central motivating factor in the implementation of new office technologies is the promise of increased productivity. In the 1940s, vendors of automatic signature machines promised to increase office production from 25 to 300 percent. Later, the promoters of electric typewriters promised that a clerk could prepare 700 premium notices a day whereas she could prepare only 600 on a standard typewriter. Some argued that a *speed-feed,* which automatically inserted and removed carbon paper from a typewriter, would enable a clerk to prepare 75 bills an hour instead of 25. The use of dictating machines was promoted as cutting a secretary's letter-preparation time in half. Several early studies on the impacts of office automation indicated that secretarial output could increase from 25 to 150 percent when word-processing equipment was used.

National statistical estimates of productivity during the period from 1968 to 1978 indicate only a 4 percent increase in efficiency for office workers. While the office technologies of this period include "primitive" machines like magnetic tape or card typewriters, they were sold as high-leverage advances when they were the best available products. In addition, this was a period in which many medium and large organizations were rapidly expanding their automation of record-keeping applications and moving some on-line. Despite substantial investments in office automation in the 1970s, overall office productivity did not increase so dramatically that national productivity measures also sharply increased. Aggregate national data cannot prove that new office technologies have had little or much impact, but they can be suggestive. Because many organizations have been continually investing in automated office equipment, the 4 percent national growth rate in productivity tempers our expectations significantly.

Whether or not information technologies generate real productivity gains for organizations is influenced by elements of work life beyond machine capabilities. Real gains can often be measured for particular tasks by focusing on one task aided by a new technology under controlled conditions. However, in real workplaces, these single-task gains do not directly translate into proportional gains in office productivity. Most office workers, aside from data entry clerks, perform several tasks. A secretary may answer phone calls, photocopy, type a document on a word processor, file, or communicate with a superior. Replacing one word processor with a "better" one may simplify some word processing tasks so that frustrating formatting problems become easier to handle. If a secretary spends 10 percent of the day formatting text, and the productivity gain is expected to be 50 percent, that translates into a much smaller true gain in overall productivity.

When real gains can be measured, people often find new tasks that exploit an increase in working capacity. Photocopiers produce more copies, text-processing machines produce more manuscript versions, and computer-based accounting systems provide more detailed accounting data. Some of these additional uses help multiply organizational products; others add little. However, the process of exploiting these "labor saving" devices is not always by saving direct labor. Rather, one adds phantom automated clerks who carry out the added work. Consequently, "extra" working capacity does not necessarily translate into cost reduction or even overall increases in effectiveness. When format changes are easier, document authors often request additional changes in a particular document. When the number of requested changes increase, these can reduce any time savings that might have accrued from using a new text formatter. Even so, many people like the feel of working with powerful tools.

Cost Savings

A second motivating factor in the implementation of new office technologies is the promise of cost savings through reduced staff. When business computers first appeared, promoters promised that machines would replace people who engaged in routine activities by simply mimicking their tasks. Computer applications would be more efficient than an office worker, and office staff could be reduced by 50 percent. Since clerical work is deemed the most repetitive and routine work done in an office, it should be the easiest to automate. However, those who have examined clerical work very closely find that many clerical tasks hinge on social judgments and negotiations that are difficult to specify, let alone automate.

Vendors do not simply sell equipment; they also sell visions about the organizational worlds within which equipment can, and perhaps should, be used. In the 1970s, IBM presented one vision of the office of the future based on a word processing plan. In its most rigid form the plan projected that in the office of 1985, individual secretaries for managers would be entirely eliminated. Instead, several managers would share services from a pool. Few organizations adopted this vision literally, but IBM was able to articulate a persuasive image of office life without secretaries.

Although there was a trend toward establishing word processing pools or centers in large offices, there has been no marked trend toward decreasing the numbers of secretaries in the labor force. In fact, the scale of clerical work in the U.S. economy has increased. From 1970 to 1980, the number of secretaries in offices increased 44 percent from 2.7 million to 3.9 million. In this same period, overall employment rose about 22 percent from 79.7 million to 97.3 million. This larger growth in secretarial work may indicate that many small and medium-sized organizations did not implement a pooled plan like IBM's proposal in the 1970s. Many organizations have found that centralized word processing alone does not effectively provide a full array of secretarial typing services. Many larger organizations today operate with a mixed mode of centralized word processing pools and decentralized office operations in which secretaries use word processing equipment in their local environment.

Cost savings by computers that replace people have been realized in special cases. For example, many telephone operators have been replaced by automatic switching devices. According to the Bureau of Labor Statistics (BLS), employment for telephone operators declined 20 percent from 397,000 jobs in 1970 to 316,000 jobs in 1980. In the remaining jobs, telephone operators use much more highly automated equipment than their counterparts 20 years ago.

The case of banking is more complex. Automated machines enable bankers to reduce the number of tellers. But between 1970 and 1980, the number of bank tellers actually rose from 288,000 to 531,000. The diffusion of automated teller machines and the displacement of some tellers is a phenomenon of the middle 1980s, not the 1970s. It appears that many banks are using automated teller machines to provide a narrow array of services continually. (Sometimes regular branch hours are left unchanged; sometimes they are reduced; at other times, branches are closed.)

Some clerical jobs have also been created through computerization. The number of computer and peripheral-equipment operators has increased about 314 percent from 126,000 to 522,000 between 1972 and 1980. Between 1972 and 1980, clerical workers increased from 14.2 million to 18.1 million. According to the BLS, these trends will continue. Over 50 percent of the 20 million new jobs projected to 1990 may be white collar jobs: managerial, professional, technical, sales, and clerical. This BLS forecast indicates that clerical work will be the fastest growing occupation and may account for about 50 percent of the new white collar jobs added by 1990.

Today, most organizations with more than $1 million per year in operating revenue have some kind of computing capabilities. Yet, the questions of office productivity and cost savings remain despite the introduction of new technologies.

Fascination with New Technologies

Many managers who work with information technologies change their sentiments from fascination, to routinization, and finally to disappointment. Fueled by vendors' promotional talk, managers become fascinated with new technologies and use the promised gains in office productivity to rationalize expensive equipment purchases. Often their expectations overestimate actual gains, even though the new technologies provide some actual value and are usually adopted. As a new technology becomes embedded in office life, equipment use becomes routine, and the limitations of new equipment become apparent. As managers become aware of better equipment in the marketplace, they become disenchanted with their original acquisitions.

These patterns can be illustrated by the example of a university department. Letter-quality printers were purchased in the early 1970s for use by faculty, students, and staff. At first, the ability to obtain good letter-quality manuscripts with a few selected daisy-wheel fonts intrigued the users. However, by the mid-1970s, faculty saw technical reports from colleagues in other departments on graphic printers that supported a wide variety of fonts and columned text. The letter-quality printers lost their appeal, and the research staff felt deprived until the department adopted laser printers in the early 1980s. Most secretaries accepted both letter-quality printers and laser printers with greater equanimity; they were less excited by the new technologies and not disappointed by the absence of graphic or laser printers. Their sentiments were more closely tied to the variety of their tasks, their relations with particular faculty, and the severe time pressures they sometimes felt.

The research staff also developed expectations for new word processing equipment that were not shared by the secretaries. The clerical workers evaluate their equipment by how easy it is for them to use, rather than by its efficiency or power. Most of the research staff were pleased when the clerical staff began to use word processing equipment in the mid-1970s. Yet, over time, their expectations for the power and efficiency of text processing has increased in parallel with the state of the art. The original equipment disappoints rather than excites. However, the clerical workers fear that new equipment will be complex and problematic; they are satisfied with the status quo, which enables them to work well enough.

Users' expectations that new equipment will improve are often realized. Electronic mail systems and text processing systems of the 1980s are generally more flexible and usable than similar systems of the 1970s. However, expectations for organizational performance rise faster than new technologies can deliver. There are two kinds of slip. First, faster electronic communication can support better organizational communication under special conditions, but the two activities are not tightly linked in general. There are fundamental limits to how much an ideal electronic communication system—whether telephone, voice mail, text mail, or fax—can improve the wide variety of communications between organizational members and the others they must work with routinely. Second, these communications technologies work best when other adjunct resources are also available. One cannot send messages at the push of a button without continual access to terminals, access to computer accounts, adequate file space, adequate telephone lines, and support staff to fix problems as they arise. Most users cannot control, or even depend on, all these resources being adequately available every minute of their working days. As a consequence, those who hope that improved technologies will lead to ideal organizations are chasing a chimerical vision.

Work in Future Offices

In the next decade, the technologies of office life will become modernized again. Many offices will acquire new technologies to support word processing, electronic mail, calendars, schedulers, laser printers, local area networks, and so on. Like the offices of today, the offices of the next decade will vary substantially in the kinds of equipment in use. Some organizations will invest heavily in the newest office technologies, while others will use limited numbers of older devices. Similarly, organizations will develop highly varied working conditions, but workers will not necessarily find better jobs in the offices with the most intensive investments of state-of-the-art equipment.

Much of the public discourse about the office of the future is based on marketing scenarios. Offices are pictured as clean, well-organized, neatly furnished, paperless, sunny, and full of plants, and with a microcomputer or terminal on each desk. These pictures give us the impression of a steady, efficient, and somewhat relaxed and cooperative mode of office work. Secretaries who look like models (actually models who look like secretaries) complete the picture with their happy and eager faces. Once the new technologies are placed in offices, we do not expect more sunlight, new furniture, more plants, and everyone fashionably dressed, nor do we expect office life to be automatically more serene. Zones of conflict that exist today will remain in the offices of the future. Conflicts over pay, prerequisites, workloads, deadlines, division of labor, and the handling of flexible time, for example, will still exist. Although some new equipment may help make some people more productive, decisions on who is going to work on which task, on what schedule, and how hard, will still have to be made.

Since the current technologies have not resolved managerial concern with office productivity, the promise of these new technologies will still be framed in the language of increased office productivity and decreased costs. The treatment and organization of work embody many social choices. These are not "givens" any more than the technological choices for office work are "given."

Further Readings

See the references at the end of the chapter.

REPORT

Reinvention

Ronald E. Rice
The University of Southern California

Word processing is a socially significant communication technology for several reasons. One, it is, for many office workers, their first and most accessible contact with computers. Second, it is a computer technology devoted to text rather than number manipulation. This indicates a major shift in the role of computers and information in modern offices. Although most information workers in modern offices handle a mixture of numbers and words, mainframe computers are predominantly used for number processing. This shift to text processing and to smaller, single-user computers is a major revolution in control and use of computing in organizations. And, third, word processing is the function people are most likely to use when they buy their own personal computers. Therefore, there is a natural learning process that often starts in the office and is brought to the home. People are beginning to use computers to help them communicate via the written (and sometimes electronically transmitted) word.

How word processing is introduced into an organization and how it is managed are critical factors in the success or failure of this communication technology. The example that follows describes how one large company first approached this problem, and how it changed that approach over time as it learned more about the implementation of word processing. The Big Bank Case comes from a sample of 200 organizations that recently adopted word processing; the study was headed by Dr. Bonnie Johnson, now at Aetna Life Insurance Company, and funded by the National Science Foundation.

Big Bank follows the trends in information systems and management; in fact they think of themselves as leaders in electronic banking and other systems. In 1974 they formed a task force to study how they might benefit from word processing. They decided that word processing was easy to use and would reduce staff but that it was not a well-understood technology. These and other conclusions were based on many kinds of studies. At that time, word processing units were too expensive to be assigned to individual secretaries, so the task force set up a center, which continued until 1979. A systematic procurement process was established whereby the center's administrators decided upon a single vendor to insure compatibility and to encourage better sales agreements.

As the center grew, Big Bank set up a word processing department that managed five centers. But the centers became too large, leading to diminishing returns. An effectiveness study concluded that output per operator dropped dramatically when there were more than 13 operators in any one center. Also, users of the centers began to want their own equipment. So the core management of word processing joined with a systems research group and an acquisition group to form a central planning group (CPG). Now they no longer manage word processing centers; users run their own centers. As one of the managers said, "Key users said, 'We want our own systems.' And others threatened to set up their own. We were so busy planning center operations that we had a hard time seeing what was happening. Planning suffered because of immediate demands."

It is important to understand that the mission of word processing administration changed from centralized control to decentralized control. "Centers should be the users' responsibility—the responsibility should be squarely on the managers of the department."

Now CPG provides information about how to determine the kind of equipment that is best for an operation. It wholesales word processing and office automation technology to the rest of Big Bank. They have developed workbooks to help offices plan their needs. These workbooks require detailed estimates of daily page output and information work. CPG decides on the appropriate equipment from its standard vendor line and delivers and maintains the equipment. Although this seems stringent, it is fairly easy to justify acquiring low-end word processing—easy enough to have a terminal for everyone to access. Word processing equipment is becoming the terminal access for Big Bank's electronic mail system.

In fact, most offices order an upgrade after about six months; this indicates that word processing is seen as effective. Note that effectiveness is now measured by accomplishment of organizational goals rather than by number of lines typed. This signals a shift from centralized planning, rational cost-benefit studies, and top-down implementation. Users want substantial control over how technology impact is measured. Like those in other organizations, users at Big Bank seem to be saying, "We'll know if the technology is doing us any good." The narrow rationality of experts from a central planning unit is seen as irrelevant at best, perhaps harmful. In tug-of-wars for resources at executive conference rooms, users exercise more power. As one CPG staff member says, "If we developed uses here and sent them out, it would be the kiss of death for us. The innovation must be user-oriented and user-driven." The CPG philosophy, evolved from a time of strict controls and rational procedures, has become today, "The user is right."

This and other cases in our study lead to several principles:

1. Evolution rather than rational planning characterizes word processing history. This history is more obvious when word processing must diffuse through an organization by means of its acceptance by users.

2. Managing word processing must be learned. Word processing was a new kind of technology for many organizations. Metaphors of typing and data processing proved to be inadequate to understanding word processing and how to manage it.

3. Conflicts between users and central providers of word processing services shaped implementation. Efficient word processing requires skilled professionals, but users tend to prefer to control the services, even if they do it less efficiently.

4. Individuals, most notably supervisors and executives, shaped the implementation of word processing.

Good managers understood the potential of the technology early and worked diligently to sell its benefits to skeptics; they put considerable effort into negotiating relationships between word processing and authors or other departments.

It may very well be that word processing, because of its unique attributes and its importance to organizational text management, is the first major communication technology since the copying machine to affect significantly how organizations go about communicating. Clearly, new management approaches are needed to meet this challenge.

Readings

Johnson, B. and R.E. Rice. *Managing Organizational Innovation: The Evolution from Word Processing to Office Information Systems*. New York: Columbia University Press, in press.

Johnson, B. and R.E. Rice. "Reinvention in the Innovation Process: The Case of Word Processing." In Rice, R.E. and Associates, *The New Media: Communication, Research and Technology,* Chapter 7. Beverly Hills, Calif.: Sage, 1984.

Rice, R.E. and J. Bair. "New Organizational Media and Productivity." In Rice, R.E. and Associates, *The New Media: Communication, Research and Technology,* Chapter 8. Beverly Hills, Calif.: Sage, 1984.

REPORT

The Patent Office Case

Paul A. Strassmann
Xerox Corporation

The test program in the office of patent attorneys is a good example of how to organize office-of-the-future projects while at the same time checking up on the economic results. This case deals with two separate groups, each doing the identical kind of work. The test group consisted of three secretaries and four attorneys, all equipped with advanced text-processing terminals that permitted easy interconnection among all participants. Text and files could be shared with ease. The test, which required participants to keep track of their time, was run for a period of six months.

The control group was also made up of three secretaries and four attorneys. They had already achieved good efficiency by means of three electronic-memory typewriters, each operated by one of the secretaries. Both the control and the test group had equivalent office conveniences, such as telephones with advanced features, and excellent copying, microfilm, and dictating equipment. Working conditions were identical.

The new equipment was made available to the test group in a manner that illustrates the proper general approach to starting up an advanced automated office:

1. Identical workstations with full-page visual displays were installed for the secretaries as well as the attorneys. These workstations were placed near the desk where each person worked. The equipment had enough local and removable memory for each individual to regard it as his or her own. The software permitted all participants to give their devices limited but nevertheless important individual characteristics for handling office communications.

2. Shared facilities, such as expensive printers and large electronic files, were easily accessible. Of particular value was a high-quality laser

printer which performed the dual function of generating book-quality paper text and of acting as a remote-access copying machine. In addition, a high-performance communications network made it possible not only to share all the electronic files within the site, but also to connect easily with all other workstations and electronic files, including an office 200 miles away.

The test explored the effects of geographic separation on cooperative work. Since patent and licensing activities are needed wherever research and development staffs are located, an important aspect of the test was an evaluation of whether office automation encouraged the continuation of geographic decentralization. If it did not, locating the attorneys at a single location might be necessary to achieve improved productivity.

It was important to limit the scope of the activity to be studied before the project began. Otherwise data gathering, data evaluation, and ability to manage the program would quickly get out of control. The focus was on five critical elements of work, leaving lesser activities for follow-up experiments. The key elements were communication and coordination, document creation and editing, filing and retrieval, legal research, and time management and scheduling.

Because a large amount of coordination was necessary between the two sites, the delay involved in passing documents back and forth turned out to be crucial to the entire experiment. The use of electronic mail, compared with mail sent by courier, improved information handling (measured by time elapsed) between sending and receipt of documents. The average document delivery time for the control group was 21 hours. The average delivery time for the test group was less than one hour. Both the average delivery time and the predictability of arrival improved. Between 8 and 46 hours prevailed for the control group. In contrast, delays for the test group never exceeded 2 1/2 hours.

From a workflow standpoint, improved predictability in the delivery of messages has unusually favorable effects on information systems involving sequential work. For example, the test group tended to spend much less time on the phone following up on the progress of projects. Quick turnaround of documents also made it possible to reduce dependence on the phone or on frequent face-to-face meetings. Incoming phone calls for the test group decreased by 45 percent, and outgoing calls decreased 27 percent. The saving of telephone time was 47 minutes per attorney per day. This gain resulted in more attention available for new work.

The statistics did not reveal any change in the frequency or length of person-to-person or group meetings. The test group observed that meetings were now used more to broaden the quality of substantive conversations with the research people, rather than to reconcile different versions of legal text.

It is also interesting to note that even though the total time spent by the secretaries on office work did not change, there were changes in the work patterns:

1. Unproductive document revision and retyping was reduced from 22 to 15 percent of total time.

2. Secretaries assumed some of the telephone activity previously handled by the attorneys, increasing their share from 1 to 12 percent of total time spent on the telephone. This was a source of improved job satisfaction because it stimulated increased social contact, which is always viewed as a more desirable form of office work than typing or filing.

3. Output, as measured in text pages, increased enormously—by 200 to 250 percent—while the time applied to typing decreased from 19 to only 15 percent of total time.

The reasons for these shifts can be explained by the following important factors:

1. Time spent on revisions of documents lowered appreciably because the attorneys personally inserted many of the lesser changes directly through their own keyboards. Revision time for a complex patent document decreased from an average of 133 minutes to 61 minutes.

2. Turnaround days for documents decreased, permitting much better use of time, reducing interruptions, and streamlining workflow.

Completion time for drafts of complex patent documents decreased from 20 to 7 days. Many more technical and legal reviews were conducted. This is why the output of pages of text reported by the test group increased so enormously without a corresponding increase in the amount of time.

The directly measurable and immediately available savings were considerable: 11 percent more hours available per attorney; 13 percent more hours available per secretary. There were numerous other, intangible benefits in terms of output quality, job satisfaction, and responsiveness to business needs. These, in the opinion of management, were sufficient to justify the entire expense for the office automation equipment.

It was gratifying to listen to the reactions of the attorneys and the secretaries in the test group. They emphasized the qualitative improvements, not the economic savings. What mattered most for the attorneys was their enhanced professional personal status. Their greatest satisfaction came from an improved ability to keep up with a large volume of cases and their clients' favorable reactions to the improved quality of legal support. Here are some of the attorneys' comments at the conclusion of the test: "The system has changed my way of doing business." "Making changes while the client is in the office and then giving him completed work is a new way of life." "The system placed more control in my hands."

The principal satisfactions for the secretaries derived from their upgraded social and economical status, achieved through increased personal contact. They were favorably motivated by the acquisition of new skills that command higher compensation. Being party to an experiment and having new opportunities to make decisions increased their job satisfaction.

Here are some of the comments reported at the conclusion of the test: "I have more time now to find out what is going on in this place." "I talk about my machine at parties. My job is more interesting. I see my way now how to upgrade my job." "Moving text around on the screen and seeing the result makes me feel that I am accomplishing something."

Perhaps the simplest way of understanding what happened to the test group can be explained by tracing the job steps performed by the control group. For instance, drafting an initial patent application required 21 separate office tasks when done within the control group. The test group shifted the creative work, including insertion of new paragraphs, to the attorneys, while relying on the secretaries to use electronic files for extracting routine text. The secretaries' role changed to a supporting, reviewing, and editing role. Total reliance on the electronic medium until the end product was printed on paper eliminated more than half the work hours. We found that in the initial patent application draft, fourteen separate job steps disappeared completely. Such changes in the patterns of workflow created favorable side effects that influenced how participants responded to the experiment. The roles of the individuals under the test conditions were sufficiently changed to warrant the conclusion that, from their point of view, a transformation of work had taken place.

Readings

Strassmann, P. *Information Payoff: The Transformation of Work in the Electronic Age*. Used by permission of the Free Press © 1985 by Paul A. Strassmann.

Topics for Research or Discussion

━━━━━ Despite its use in everyday discussions of business, the concept of "productivity" has a variety of meanings. Do some research on this topic; one place to begin is with the references to this chapter. Attempt to find differences in the meanings ascribed to productivity. Also, see if you can locate suggestions for the measurement of productivity in actual settings.

━━━━━ Somewhat to the surprise of researchers who are looking for more dramatic impacts, respondents to surveys of the business users of computer terminals or other video display technologies often complain of eye strain or neck pains, or express a fear that radiation from such equipment may be hazardous to their health. Prepare a consulting report on this topic for a major corporation where there are many terminals in use. Find out not only what the state of knowledge is on this topic, but what some managers have done to try to alleviate the problems. (TIP: Any index to a major business publication, for example, *The Wall Street Journal,* will be helpful to you. Also consult your reference librarian.)

━━━━━ See if you can gain the cooperation of a small business or a departmental unit in an organization (for example, an academic department in a university) for the purpose of analyzing the use of telephone services. What are the major functions of telephone use? To what degree do the various functions contribute to the primary needs of the business or organization? If possible, find out how much is spent on telephone services and examine this relative to the overall business or departmental expenditures. Based upon your analysis, do you think that the telephone could be used more efficiently? If so, how?

━━━━━ "Reinvention" in this volume refers to how users discover new uses for information technologies. (For example, telephone answering machines were designed to record calls while the phone was unattended, yet many people use them to screen calls or to leave "voice mail" type messages.) Give an example of reinvention, real or imaginary. Discuss how or why that reinvention may come about. Are there any further prospects for reinvention in the example you give?

━━━━━ The processes of planning and acquiring new technologies are important parts of modern management. Gain the cooperation of a manager in a local business or other type of organization (e.g., the university) where a new technology has been acquired (e.g., word processors). Conduct a brief case study of how planning was accomplished, as well as how the purchase was made (i.e., how did they do the "shopping?").

References and Further Readings

Ackoff, R. "Management Misinformation Systems." *Management Science* 2(1967):147–156.

Bair, J. "An Analysis of Organizational Productivity and the Use of Electronic Office Systems." In *Proceedings of the American Society for Information Science* 43(1980):4–9.

Bair, J. "Productivity Assessment of Office Information Systems Technology." In *Trends and Applications: 1978 Distributed Processing.* Washington, D.C.: IEEE and National Bureau of Standards, 1978, 12–22.

Bamford, H.E., Jr. "Assessing the Effect of Computer Augmentation on Staff Productivity." *Journal of the American Society for Information Science* (1979):136–142.

Bostrom, R. and J. Heinen. "MIS Problems and Failures: A Socio-Technical Perspective. Part II: The Application of Socio-Technical Theory." *Management Information Science Quarterly* (December 1977):11–28.

Bullen C.V., L. Bennett, and D. Carlson. "A Case Study of Office Workstation Use." *IBM Systems Journal* 21(1982):151–169.

Business Week. "The New Broader Gauges of Productivity." 19(1982):44B–44J.

Carlson, E. "Evaluating the Impact of Information Systems." *Management Informatics* 3(1974):56–67.

Chandler, J. "A Multiple Criteria Approach for Evaluating Information Systems." *Management Information Science Quarterly* 6(1982):61–74.

Conrath, D. "Organizational Communication Behavior: Description and Prediction." In M. Elton, W. Lucas, and D. Conrath, eds., *Evaluating New Telecommunication Services*. New York: Plenum, 1978, 425–442.

Crawford, A., Jr. "Corporate Electronic Mail—A Communication-Intensive Application of Information Technology." *Management Information Science Quarterly,* 6(1982):1–13.

Culnan, M. and J. Bair. "Human Communication Needs and Organizational Productivity: The Potential Impact of Office Automation." *Journal of the American Society for Information Science* 34(1983):218–224.

Curley, K. and P. Pyburn. "Intellectual Technologies: The Key to Improving White Collar Productivity." *Sloan Management Review* (Fall 1982):31–39.

Dordick, H. and F. Williams. *Innovative Management Using Telecommunications*. New York: John Wiley, 1986.

Downs, C. and T. Hain. "Productivity and Communication." Vol. 5, *Communication Yearbook,* edited by M. Burgoon. New Brunswick, N.J.: Transaction Books, 1982, 435–454.

Driscoll, J. "Office Automation: The Organizational Redesign of Office Work." Cambridge, Mass.: MIT Center for Information Systems, Research Report 45, 1979.

Drucker, P. *The Changing World of the Executive*. New York: Times Books, 1982.

Drucker, P. *Managing in Turbulent Times*. New York: Harper & Row, 1980.

Edwards, G.C. "Organizational Impacts of Office Automation." *Telecommunications Policy* (June 1978):128–136.

Elizur, D. and L. Guttman. "The Structure of Attitudes toward Work and Technological Change within an Organization." *Administrative Science Quarterly* 21(1976):611–622.

Ganz, J. and J. Peacock. "Office Automation and Business Communications." *Fortune* (October 5, 1981):7ff.

Giuliano, V. "The Mechnization of Office Work." *Scientific American* 247(1982):148–165.

Goldfield, R. "Achieving Greater White-Collar Productivity in the New Office." *Byte* 8(1983):154ff.

Gutek, B. "Effects of Office of the Future Technology on Users: Results of a Longitudinal Field Study." In G. Mensch, and R. Niehaus, eds., *Work, Organizations and Technological Change*. New York: Plenum, 1982.

Harkness, R. "Office Information Systems: An Overview and Agenda for Public Policy Research." *Telecommunications Policy* (June 1978):91–104.

Helmreich, R.K. "Field Study with a Computer-Based Office System." *Telecommunications Policy* (June 1982): 136–142.

Hiltz, S.R. *Online Communities: A Report of the Office of the Future*. Norwood, N.J.: Ablex, 1983.

Johnson, B. and R. Rice. *Managing Organization Innovation: The Evolution from Word Processing to Office Information Systems*. New York: Columbia University Press, 1987.

Johnson, B. and R. Rice. "Policy Implications of New Office Systems." In V. Mosco, ed., *Telecommunications Policy Handbook*. Norwood, N.J.: Ablex, 1984.

Johnson, B. and R. Rice. "Redesigning Word Processing for Productivity." In R. Vondran, ed., *Proceedings of the American Society for Information Science*. Washington, D.C.: ASIS, 1983, 187–190.

Keen, P. "Information Systems and Organizational Change." *Communications of the ACM* 24(1981):24–33.

Keen, P. and S. Morton, *Decision Support Systems*. Reading, Mass.: Addison-Wesley, 1978.

Landau, R. "Office Automation in the U.S. and Japan." *ASIS Bulletin* 9(1983):6–11.

Lodahl, T. "Cost-Benefit Concepts and Applications for Office Automation." Ithaca, N.Y.: Cornell University/Diebold Group, 1980.

Montgomery, I. and I. Benbasat. "Cost/Benefit Analysis of Computer-Based Message Systems." *Management Information Science Quarterly,* (March 1983):1–14.

National Archives and Records Service. *Office Automation in the Federal Government: A Status Report*. Washington, D.C.: U.S. Government Printing Office, 1981.

Newman, J. "Human Factors Requirements for Managerial Use of Computer Message Systems." In R. Uhlig, ed., *Computer Message Systems*. New York: North-Holland, 1981, 453–465.

Olson, M. "New Information Technology and Organizational Culture." *Management Information Science Quarterly* (1982): 71–84.

Packer, M. "Measuring the Intangible in Productivity." *Technology Review* (February/March 1983):48–57.

Panko, R. "Office Automation Needs—Studying Managerial Work." *Telecommunications Policy* (December 1981):265–272.

Panko, R. "Serving Managers and Professionals." In J. Sutton, ed., *1982 Office Automation Conference Digest*. San Francisco: American Federation of Information Processing Societies, 1982, 97–103.

Patrick, R. "Probing Productivity." *Datamation* (September 1980):207–210.

Poppel, H. "Who Needs the Office of the Future?" *Harvard Business Review* 60(1982):146–155.

Pye, R. and I. Young. "Do Current Electronic Office Systems Designers Meet User Needs?" In R. Landau, J. Bair, and J. Siegman, eds., *Emerging Office Systems*. Norwood, N.J.: Ablex, 1982, 73–94.

Shulman, A. and J. Steinmen. "Interpersonal Teleconferencing in an Organizational Context." In M. Elton, W. Lucasa, and D. Conrath, eds., *Evaluating New Telecommunication Services*. New York: Plenum, 1978, 399–424.

Strassman, P. *Information Payoff: The Transformation of Work in the Electronic Age*. New York: Free Press, 1985.

Strassmann, P. "Managing the Costs of Information." *Harvard Business Review* (September/October 1976):133–142.

Strassmann, P. "Office of the Future." *Technology Review* (1980): 54–65.

Tapscott, D. *Office Automation: A User-Driven Method*. New York: Plenum, 1982.

Tersine, R. and R. Price. "Productivity Improvement Strategies: Technological Transfer and Updating." *Journal of Systems Management* (November 1982):15–23.

Tuttle, H. "Bringing Your Productivity Improvement Efforts into Focus." *Productivity* 89(1982):68–75.

Uhlig, R., D. Farber, and J. Bair. *The Office of the Future: Communications and Computers*. New York: North-Holland, 1979.

Williams, F. and H. Dordick. *The Executive's Guide to Information Technology: How to Increase Your Competitive Edge*. New York: John Wiley, 1983.

Zisman, M. "Office Automation: Revolution or Evolution?" *Sloan Management Review* 19(1978):1–16.

Zuboff, S. "New Worlds of Computer-Mediated Work." *Harvard Business Review* 5(1982):142–152.

9

The Public Media

Public is the broadest concept of a communication context explored in this volume. In one overall respect, it comprises those media systems that disseminate messages throughout a society, the topic of this chapter. But it also reflects upon the circumstances and motives for the usage of such media, the topic of the next chapter.

TOPICAL OUTLINE

NOTES ON PUBLIC COMMUNICATION

The Concept of Public

Public communication refers to messages directed at or available to large numbers of individuals who are not in physical proximity to the communication source. Although this is close to the definition of *mass communication,* we take *public* as a more encompassing concept.

For present purposes, all communication intended for consumption by large groups of people will be called public. Within this area are the traditional mass media of print, broadcasting, and film, in which the media products are delivered widely to those who buy, subscribe, or tune in. The *mass* of mass communication was originally applied as an analogy to the 19th-century concept of mass society, in which the public was largely faceless and anonymous. However, the trend in this century has been for audiences to become increasingly identified, segmented, and catered to.

As public media alternatives have proliferated, the increases in choice make the recipient less anonymous, a process some call *demassification*. (Among the examples are electronic text or database services, from which persons request highly individual and specific information.) We have also seen the increasing focus of public media on specific audiences, both under commercial conditions of free enterprise and under political systems in which media serve the government. These various forms of communication, together with the traditional mass media, constitute our concept of public communication.

Figure 9.1 summarizes general features of the public communication process. After a brief examination of the traditional characteristics of public media, the remainder of this chapter focuses on technological change and its implications.

Media Technologies for Dissemination

Public communication requires media that allow for message distribution to large audiences. Print technology is often acknowledged as the first mass medium, although the more broadly construed concept of public communication is reflected in the publicly available art, architecture, or amphitheaters of antiquity. In our times, radio is considered to be the medium that is most attended to (in both time and numbers) by the world's publics, followed possibly by television. Between print publication and broadcasting stand film, audio, and video recording technologies, and broadcasting enhanced by coaxial cable, microwave, and satellite transmission.

Traditionally, as in the characterization of mass communication, the flow of dissemination is from a central source to a broad, undifferentiated audience, a pattern of one-to-many. Upon closer examination, however, there are notable exceptions. Films are typically viewed by members of audience groups, although the groups may shrink to living room size as film distribution via video cassette grows. Television, too, is often watched by small groups.

Current technological trends that provide for interactive public media, such as are approached by two-way cable (typically, video one way and data on the

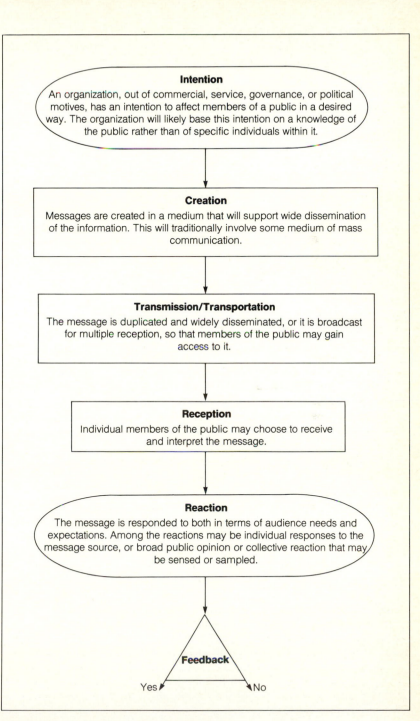

Figure 9.1 *The Public Communication Process*

return) or videotext systems, reflect a shift toward a many-from-one character in which the audience member gains the message from the source upon request. This is somewhat analogous to a subscriber requesting a magazine, only in this case the service, if not the specific message, is requested.

Broadly speaking, media technologies have served the growth of public communication not only through methods of message duplication and dissemination but also by now adding further alternatives through implementation of interactive capabilities.

A GLIMPSE AT THE WORLD'S MEDIA

We Americans usually examine media strictly from a U.S. or Western bias. In reality, the world-wide production of public media is impressive; some of the highlights of the world statistics are summarized in Table 9.1 for print media and Table 9.2 for broadcasting.

NEWSPAPERS AS TECHNOLOGY ADOPTERS

Behind the Scenes

Most members of the newspaper-reading public do not realize that the paper delivered to them each morning or evening reflects one of the most major adoptions of modern communication technologies. The newspaper may look much like one of several decades ago, but the mechanics for its production have changed. New forms of hardware and software predominate in the production or front-end system of most newspapers. The major components of electronic publishing behind-the-scenes are summarized in Table 9.3. The transition from the old technology to the new came quickly. In 1973, no computer-oriented systems could be found in any newspaper; but within five years, major papers had complete systems. For editors, this has meant enhanced responsibility by putting the entire production process into their hands. They also have greater flexibility, more time, and fewer restrictions.

In 1983, several newspapers reported experimenting with the new breed of lap computer. They reported that the portable computer was useful for reporters because of its size. With its full-sized keyboard, writing, editing, moving, and retrieving text is easy. For example, note-taking in a courtroom is facilitated. The reporter need not write notes in longhand and then type them into a story or dictate the story over the phone. The original keystroke is captured and utilized. This meant the traditional phone-in story would become obsolete. There was no more need to involve separate individuals—one to dictate and one to receive dictation—to produce a story. Another application for portables has been in the numerous daily short stories that can be so time-consuming: sentencing and pleas in the clerk of court's office, burglaries, and beatings and

Table 9.1 *Summary Statistics on Print Media in the World*

DISTRIBUTION OF BOOK PRODUCTION
(number of titles by continents and major areas; estimated percentage 1975, 1980, and 1982)

1975——Europe-USSR 60.3; N. America 16.2; Asia 15.1; Latin America 5.1; Africa 1.4; Oceania 1.0; Arab States 0.9

1980——Europe-USSR 55.9; Asia 19.5; N. America 15.9; Latin America 4.7; Oceania 1.7; Africa 1.3; Arab States 1.0

1982——Europe-USSR 55.1; Asia 19.7; N. America 15.3; Latin America 5.9; Oceania 1.5; Africa 1.4; Arab States 1.1

DISTRIBUTION OF DAILY GENERAL INTEREST NEWSPAPERS BY CONTINENTS AND MAJOR AREAS
(estimated percentage 1975 and 1982)

1975——Europe-USSR 53.0; Asia 21.8; N. America 16.2; Latin America 5.6; Oceania 1.7; Africa 1.0; Arab States 0.7

1982——Europe-USSR 45.6; Asia 31.4; N. America 13.2; Latin America 6.4; Oceania 1.2; Arab States 1.2; Africa 1.0

DAILY GENERAL INTEREST NEWSPAPERS, 1982
(by country, number, and circulation)

Egypt, 10, 3484 (thousands); Nigeria, 15, 510; Canada, 120, 5570; Mexico, 374, 10,212; U.S., 1710, 62,415; Argentina, 191, 3343; Peru, 68, 1490;
China, 53, 33,654; India, 1087, 13,033 (1979); Japan, 154, 68,142; W. Germany, 368, 25,103; France, 90, 10,332; U.K., 113, 23,472; USSR, 772, 109,306

PERIODICALS OTHER THAN GENERAL INTEREST NEWSPAPERS, 1982
(by country, number, and circulation)

Egypt, 187, 1262 (thousands); Canada, 1384, 56,169; Mexico, 1964, no circulation available (NCA); U.S., 59,609, NCA; Peru, 507, NCA; China, 3100, 138,852; Japan, 1640, 25,604; France, 13,716, 183,379; USSR, 5358, 4,258,182

NUMBER AND CIRCULATION OF DAILY NEWSPAPERS
(by continent or major area, 1982)

Africa, 120 dailies, 5 million circulation; N. America, 1830, 68; Latin America, 1200, 33; Arab States, 110, 6; Asia, 2430, 162; Oceania 110, 6

From *Statistical Yearbook 1984.* Paris: United Nations Educational, Scientific and Cultural Organization, 1984.

Table 9.2 *Summary Statistics on Broadcast Media in the World*

DISTRIBUTION OF RADIO RECEIVERS BY CONTINENTS AND MAJOR AREAS
(estimated percentage 1975, 1980, and 1982)

1975——N. America 45.4; Europe-USSR 29.8; Asia 11.0; Latin America 8.7; Arab States 1.8; Africa 1.8; Oceania 1.4

1980——N. America 41.5; Europe-USSR 28.9; Asia 14.3; Latin America 9.3; Africa 2.2; Arab States 2.1; Oceania 1.7

1982——N. America 38.0; Europe-USSR 26.8; Asia 19.9; Latin America 9.2; Africa 2.4; Arab States 2.1; Oceania 1.6

DISTRIBUTION OF TELEVISION RECEIVERS BY CONTINENTS AND MAJOR AREAS
(estimated percentage 1975, 1980, and 1982)

1975——Europe-USSR 43.0; N. America 33.8; Asia 14.0; Latin America 6.9; Oceania 1.4; Arab States 0.9; Africa 0.2

1980——Europe-USSR 44.2; N. America 29.0; Asia 15.8; Latin America 7.4; Arab States 1.6; Oceania 1.3; Africa 0.7

1982——Europe-USSR 42.4; N. America 28.5; Asia 17.7; Latin America 7.6; Arab States 1.8; Oceania 1.3; Africa 0.7

NUMBER OF RADIO BROADCASTING TRANSMITTERS
(by continent or major area)

1980——Africa 680; Asia 2810; Arab States 360; N. America 10,100; Latin America 4570; Oceania 430

1981——Africa 720; Asia 2850; Arab States 380; N. America 10,400; Latin America 4800; Oceania 450

NUMBER OF RADIO BROADCASTING RECEIVERS
(by continent or major area)

1980——Africa 27 (million); Asia 235; Arab States 25; N. America 496; L. America 107; Oceania 19

1982——Africa 33; Asia 270; Arab States 29; N. America 514; L. America 124; Oceania 22

NUMBER OF TELEVISION RECEIVERS
(in millions, by continent or major area)

1980——Africa 3.5; Asia 90; Arab States 8.4; N. America 153; L. America 39; Oceania 6.5

1982——Africa 3.8; Asia 100; Arab States 10.2; N. America 161; L. America 43; Oceania 7.5

Table 9.2 *(continued)*

RADIO BROADCASTING RECEIVERS, 1982
(by country and number)

Egypt 7000 (thousands); Nigeria 6600; Canada 18,668; Mexico 21,300; U.S. 495,000; Argentina 21,200; Brazil 45,000; China 65,000; India 40,000; Japan 82,400; France 46,300; W. Germany 24,158; U.K. 55,000; USSR 136,000

NUMBER OF TELEVISION RECEIVERS, 1982
(by country and number)

Egypt 1850 (thousands); Morocco 845; Nigeria 455; S. Africa 2200; Canada 11,316; Mexico 8000; U.S. 150,000; Argentina 5900; Brazil 15,500; Colombia 26,000; China 6000; Japan 66,342; France 20,000; W. Germany 21,834; U.K. 25,500; USSR 83,000

From Statistical Yearbook 1984. Paris: United Nations Educational, Scientific and Cultural Organization, 1984.

---

muggings from the police report stacks. The device can even be taken home at night. If anything newsworthy happens, the reporter can write a story on the spot and send it into the newspaper from home.

Effects upon Writing and Editing

A content analysis study (Randall, 1979) hypothesized that newspapers using electronic editing systems have shorter sentences and fewer errors in spelling, punctuation, sentence construction, hyphenation, and typography than newspapers not using electronic editing systems. Selecting the *Charlotte* (N.C.) *Observer* for study, the author found his hypothesis largely supported (with the exception of sentence length). The author warned against generalization—

Table 9.3 *Modern Newspaper Technologies*

Input devices: optical character readers, video display terminals, on-line keyboards, and off-line keyboards

Storage devices: central computers, floppy disks, punch-paper tapes, and distribution units

Transmission devices: microwave, telephone network, coaxial cable, communications satellites

Processing devices: central computers and minicomputers

Editing and makeup devices: video display terminals and electronic layout boards

Output devices: phototypesetters, on-line typewriters, and line printers

---

because of unequal electronic systems and staff changes—to other newspapers, however.

There has also been research into the effects on editing skills, as measured in terms of speed and accuracy for both mechanical and judgmental editing abilities (Shipley and Gentry, 1981). Copy editors with at least six months' experience with video display terminals (VDT) were tested at four randomly chosen and geographically dispersed newspapers. They found few statistically significant differences in time or errors for pencil-versus-VDT editing. But trends in their data indicated that VDT editing is slower but more accurate than pencil editing. Some editors were faster or more accurate than others, but speed and number of errors were not found to be related. Only age and circulation size of the newspaper were found to be associated with differences in editing speed.

Satellites

Replacing slower mail service, satellite transmission now gives the newspaper industry a practical and almost spontaneous method for sending information to any newspaper across the country. Unlike other communication industries, newspapers did not begin to make widespread use of satellite technology until 1979 when government regulation requiring expensive licensure for receive-only satellite dishes was lifted. The two major wire services—the Associated Press and United Press International—have moved rapidly to install receiving stations, thus allowing them independence from and considerable savings over transmission through telephone systems. This technology also made it possible to transmit newspapers into subscribers' homes through television. More efficient communication within franchise businesses and videoconferencing also are possible. The most significant advantages of using satellites for news gathering will be realized if technology continues to make satellite communications more portable and less expensive.

In the late 1970s, the *Wall Street Journal* faced the dilemma of increasing demand but insufficient production capacity, forcing it to limit circulation and advertising growth. But in 1982 it opened a satellite printing plant, completing a construction program that enabled it to print and distribute 2.2 million copies using the largest and most comprehensive satellite transmission system in the newspaper industry. The process begins at the *Journal's* New York offices, where editorial material is entered into computers linked to the Chicopee, Mass., composition and production plant. There the editorial is combined with national advertising material and sent by microwave to the *Journal's* South Brunswick, N.J., plant. From there, the contents of the newspaper are transmitted via the Westar V satellite to the other composition/production plants, grouped into four regions: eastern, midwestern, southwestern, and western.

In 1984, the *Christian Science Monitor* also converted one of its plants to satellite technology. Paper officials, however, feel they covered new ground in the subject. Although satellite facsimile transmission is common to other nationally produced newspapers, *Monitor* officials feared the reputed costliness of transmitting the enormous quantities of data found in the newsprint. Such transmission has been accomplished via a transmission band capable of carrying 1.5 million bits of information per second. The *Monitor's* system uses a narrower bandwidth of 56,000 bits per second capacity. They feel this has a direct bearing on transmission cost.

Satellite communications seems to be a perfect match for prepress-to-press transmission, with numerous advantages over terrestrial methods. Logistics is one reason. *The Christian Science Monitor* had committed to off-site printing as part of a broad plan to strengthen the already highly regarded publication. Arrangements were made with the telephone company for dedicated land lines. But just 60 days before the target date for the off-site printing was to begin, the telephone company told the publisher that the network was logistically impossible. Shortly afterward, newspaper officials heard about the network that Modulation Associates had established for the Olympics. Negotiations were made, and a portable uplink was flown to Boston while the newspaper's permanent uplink could be put on-line.

Affordability is another satellite advantage. A satellite link typically is less expensive than its terrestrial counterpart. Moreover, the customer owns the network, eliminating constant user fees. Additional savings come from possible reduction in interface hardware costs. Recent state-of-the-art advances in satellite hardware design promote further advantages of flexibility and expandability. The introduction of the low-power, all-solid-state satellite uplink is an example. Low power means low cost, and the solid-state components eliminate breakable tubes and dangerous voltages. The satellite receiver also has been improved in the last few years. Not only has inter-receiver compatibility been increased, but because the receiver is designed for multichannel capability, one receiver can serve multiple uses.

Facsimile

The transmission of newspaper pages by **facsimile** began with experiments in Japan in 1959 (Russell, 1981). At that time it took 30 minutes to transmit one broadsheet page of mediocre quality (300 lines/inch). The first transmission of a newspaper page by satellite was in 1967 when the front page of the *London Daily Express* was transmitted from London to San Juan as part of a demonstration for the American Newspaper Publishers Association convention in Puerto Rico. Page facsimile transmission can have several effects. If there is a demand—and hence the economic viability—a newspaper can now distribute anywhere in the world.

There also is the potential for the mobile print shop. For example, *The London Daily Telegraph* launched a special edition on board the Queen Elizabeth 2 liner from 1969 to 1976. Minor transmission garble, slowness, and picture-transmission problems indicated the need for a better system, which was developed in the 1980s. This rapid-scanning system was linked to geostationary satellites.

Newspapers also have used facsimile transmission to deal with geographic problems. For example, the Norwegian newspaper *Aftenposten* installed a fiber optics link to transmit pages from the Oslo head office to a suburban production plant—just six miles away—to circumvent bad traffic and weather.

WHAT'S AHEAD FOR NEWSPAPERS?

Was McLuhan Right?

It has been well over two decades since communications pundit Marshall McLuhan (1962) warned that our media were going electronic and that in our lifetimes we would witness the steady demise of the printed word.

From the vantage point of the late 1980s it appears that McLuhan was only partly right. For one, you are not reading this chapter in some new electronic form but on the printed page. However, there has been substantial change behind the scenes. This book was written on a computer, edited on a different one, then set in type by still another one. Most words that you read today, especially in newspapers, are produced in their printed format by technologies—computers, editing programs, and telecommunications systems— that have already far outstripped even McLuhan's most venturesome visions. To date, the electronics revolution has advanced the art of paper-oriented publishing, not replaced it.

Moreover, there are signs of still further change. If the production of newspapers involves so much manipulation of text in electronic form, why bother printing it for delivery to the reader? Why not send electronic text over TV broadcasts, cable systems, or via telephone lines to home computers, to be read on your TV or computer screen? If you did want a paper version, why not print selected stories in your home on a machine hardly more complex than your electric typewriter?

Electronic Delivery

The prospect of electronic delivery is the drama of the new text services. Are we about to bid goodbye to trucks, neighborhood newspaper carriers, and newsstands? Will you read your morning headlines on a computer? In short, will electronic distribution of your newspaper replace the traditional transportation-based system? The answer to this question is now shaping up on the horizon.

One part of the answer is that many readers see little advantage to browsing through the news by selecting stories from a menu on a TV or computer screen compared with scanning the printed page. Time, Inc., spent undisclosed millions to test whether there would be an electronic text market in the home, only to withdraw from that business for the present. The New York Times Co. recently disbanded its electronic text services in favor of a third-party (Mead Data Central) publisher. And the *Washington Post* pages available through the

CompuServe (H & R Block Co.) text service are not among the most popular of the latter company's selections.

Experiences with Dow Jones News/Retrieval Service somewhat confirm the foregoing generalization but also add an interesting counterpoint. Although you can read the *Wall Street Journal* daily via your personal computer and a telephone link, there appears to be little evidence that this has cut into the circulation of the traditional paper edition (so said William L. Dunn, president of that service, when queried at a conference on electronic publishing). Other than the lack of enthusiasm about electronic newspage formats, a reason for noncompetitiveness may be that to date you cannot get access to the electronic edition of the *Wall Street Journal* any earlier each day than to the paper edition. Thus there is no time advantage.

But the Dow Jones electronic information services are profitable; presumably this is because there is a time and convenience advantage in obtaining market data and research information on demand via a computer terminal. Moreover, there are conveniences in being able to interact with the services, where, for example, the answer to one question may lead you to ask another. Dow Jones is also pursuing another potentially profitable application of its retrieval services—that of selling computer programs that allow subscribers to load market information directly into their computers and then to analyze it according to personal specifications. These services are all more akin to information retrieval from files (or databases) than the act of browsing through a newspaper.

The popularity of specific information retrieval over scanning or browsing through a wide variety of content has also been revealed in the experiences with the British PresTel system. Originally, PresTel (initiated by the postal and telecommunications service) was to be a wide-ranging electronic publishing system whereby subscribers could request anything from the day's news to travel reservations. After a near decade of service, PresTel has evolved to much more of a specialized set of database services than a mass information system. Subscribers are more prone to be in search of information or services when time, rapid access, and interaction are valuable.

New Entries in the Publishing Business

Also arriving on the scene are companies with no relation to the earlier publishing business. For example, if banks find that they can process electronic checks at a tenth of the cost of paper checks, it is obviously to their advantage to encourage customers to use new on-line services. With all the success that banks have had in encouraging us to use automatic teller machines, it is not too much of a step to having us use our home computers for checking or transferring funds among our accounts. This adds several more twists to the growth of electronic publishing. First, if consumers are already banking via their home terminals (perhaps furnished by the bank), then it is an easy step to use that terminal for other purposes. Also, in their rush to promote financial services, banks may be offering premiums in the form of electronic mail, home shopping, and even news headlines.

There is also the question of whether text services could ever attract the magnitude of advertising dollars that support the newspaper industry. Experiences such as the foregoing suggest that electronic information services, because of their lack of attraction as a mass medium, may never achieve the type and numbers of circulation that are required by major advertisers. On the other hand, the data retrieval nature of these systems lends itself very handily to classified advertising, and an interactive service can offer the opportunity for readers to respond directly to ads.

A Conservative View of Change

So where does all of this take the future of the newspaper? As was the case with television's entrance into the news business, you may get more of your headline information from electronic text services. Or if you want to see if some specific topic is in the day's or week's news, you might wish to use your computer to search for that information in a newspaper's electronic files. But for skimming over the day's news, or reading longer interpretative stories, your traditional newspaper will probably survive in its present form—perhaps slimmer, with better pictures and graphics, with color, with more interpretative features, and with increased local coverage. This conclusion fits the long-recognized generalization that new media do not usually replace older ones; instead, media evolve to fit new patterns of use.

CHANGES IN BROADCAST AND VIDEO MEDIA

The late 1970s were a preview of the many viewing channels that cable TV could bring. The most popular was Home Box Office (HBO), which offers recent movies without commercials—all for a monthly fee. Another early satellite channel was Nickelodeon, which provided programs to entertain and educate children in the manner of *Sesame Street*. A flood of devices—multichannel cable systems, video recorders, and disk players—spread rapidly and provided competition for traditional network TV fare.

Pay TV networks seemed to come into their own in the late 1970s, riding piggyback on cable systems. At that time they were dominated by HBO, possessor of 70 percent of the pay-cable market in 1979. HBO's closest competitor was Showtime, with about 21 percent. Subscription TV (STV) attracted headlines in areas without cable. STV has been described as the broadcast version of pay TV, but physical and potential regulatory limitations on its transmissions have limited its growth to certain markets (Bernstein, 1979).

The appearance of *superstations* was also a controversial phenomenon at the time. These stations were created by bouncing the signals of a local broadcaster off a satellite to transmit them, for a fee, to cable operators across the country. Broadcasters in the invaded markets were upset because the outsiders—Atlanta's WTCG, New York's WOR, Chicago's WGN, and Oakland's KTVU—sometimes brought in the same programs that the local stations carried. Videotape recorders and video disks promised to bring even further changes to the industry, but the former has been the more successful of the two. In the first two years they were on the market, some 600,000 VCRs were sold while only 2000 video-disk players were bought.

This same period was also laden with innovations for the broadcasting industry, from microprocessor chips to helicopters equipped with cameras and microwave receivers. Many gains made by broadcasters came via application of digital techniques, especially computers and microprocessors, which are expected to lead the way in the 1980s. Videotape recorders, for example, are generally believed to have advanced as far as they can in analog; the next step is digital. Satellite technology became more prominent in broadcasting when The Communications Satellite Corporation (COMSAT) announced plans in the last decade to start direct-to-home subscription TV service via satellite. Local TV stations got more into the act by purchasing earth stations. Radio stations also got involved with satellite distribution because it allowed them to send out simultaneously a number of different circuits, allowing for increased flexibility in the regionalization of programming and commercials.

When the FCC deregulated earth stations in 1979, it spawned an entire new medium. Cloned from cable systems, **SMATV** (satellite master antenna TV) systems operate on private property such as apartment buildings, multiunit condos, and mobile home parks (see *Broadcasting,* 1982). But instead of procuring franchises, SMATV operators sign contracts with the property owners, allowing them to bring cable TV to homes on the property. This usually is done by simply hooking up an earth station aimed at a satellite to a master antenna system.

Another sister technology to cable TV is **multipoint distribution service** (MDS). MDS is a microwave common-carrier broadcast technology that can disseminate television signals within a 20-mile radius. MDS signals principally are beamed to hotels, motels, and business establishments, although operators have begun to solicit home hook-ups for pay TV (McCavitt, 1983).

REPORT

Videotext in Japan

Liz Greenberger, with the assistance of Alexandra Bash
Council for Technology and the Individual

Japan has long been characterized as an imitator of Western technology. Masters at dissecting and improving upon American designs, the Japanese have spent a good part of their R&D effort catching up, especially in fields such as telecommunications. Today, they appear to have arrived. Not only has Japan proclaimed to the world that it intends to shed its copycat image, it has announced bold plans to pioneer in systems based on knowledge and information. In fact, it is being called

> . . . the first nation to act consciously upon the realization that the . . . wealth of nations which depended upon land, labor, and capital during its agricultural and industrial phases . . . will come in the future to depend upon information, knowledge and intelligence. (Feigenbaum and McCorduck, 1983)

This quote refers to a program in Japan, called the Fifth Generation Computer project, to develop a new breed of superfast, superintelligent computers. Underlying national policy emphasizing dominance in information processing seems to be the foundation for all Japanese R&D in computers and communications today.

It is with this in mind that we ask whether the Japanese are likely to become pioneers in the field of videotex as well.

Videotex dates back to work in Great Britain in the second half of the 1970s. Although not a well-defined concept, videotex is most often viewed as a system "for the widespread dissemination of textual and graphic information by wholly electronic means for display on low-cost terminals (often suitably equipped television receivers), under selective control of the recipient, and using control procedures easily understood by untrained users" (Tydeman, 1982).

Many countries now have operating videotex systems. The primary videotex system in Japan is called the CAPTAIN (Character and Pattern Telephone Access Information Network) system. Designed to use standard telephone circuits and an adaptor, it hooks the television receiver to the phone line, enabling precoded messages in frame form to be transmitted from the CAPTAIN center.

A pilot service began on December 25, 1979. During the first three months of the experiment, 1000 monitors in the Tokyo metropolitan area were connected to an information bank that was capable of storing and transmitting about 100,000 pages of data. Eight hundred of the 1000 terminals were in private homes. In the second phase of the trial, the database was augmented by a larger number of information providers, and 500 additional receiving terminals were installed in Japanese businesses.

The trial began with 18,600 frames of information. By April 1980, it had grown to over 82,000 frames. By early 1982, CAPTAIN was serving close to 2000 terminals: 1200 for residential users, 500 for business users, 200 for information providers, and 100 for public access and system development. The initial 165 information providers had expanded to 199, and the database mushroomed to 204,310 frames.

The service was provided free of charge during the experiment, but 30 percent of the users indicated they would be willing to pay more than 3000 yen (approximately $13) per month to receive it (Sigel, 1983).

Full-scale commercial CAPTAIN service officially began on November 30, 1984. Information providers can use the videotex service to send pictures and text over the phone lines, or they can connect their own computer to the CAPTAIN center. The center currently provides an interactive videotex communication system, a database service, terminal facilities, closed user group systems for business applications, and transaction-based systems for residential users.

The most successful of the enhanced services is the closed user group (C.U.G.) and transaction services. A C.U.G. service allows information providers to offer specific information to a limited group of selected members. It is often used as a communications network within a company or between headquarters and satellite offices.

The transaction service, also known as the order entry service, handles reservations for theater, trains, airlines, and hotels. It also processes shopping orders and banking transactions on a 24-hour-a-day, 7-day-a-week basis. Recent reports indicate that the most common and successful use is by financial service companies for providing financial data to businesses (Doe, 1984).

The residential videotex development in Japan has been distinct from that in other countries because it has been shaped by the intricacies of the Japanese language. Japanese is a combination of over 3000 kanji characters plus phonetic hiragana and katakana symbols as well as numbers and roman letters.

Transmitting Japanese ideographs electromechanically has been a serious design challenge. For videotex in Japan, it has meant a much greater emphasis on graphics. Graphics require more storage and mean more expensive receiving terminals for display. All of this adds up to more computing power and a more costly service to operate. Costly services are not usually attractive to mass audiences, especially if the information they provide can be had by less expensive means. Although the Japanese may be leaders in the technical transmission of complex videotex graphics, they have as yet been unable to find the "trigger service" (Hopper, 1985) or combination of services that would motivate a large number of consumers to make the financial investment necessary for CAPTAIN's success.

Standardization has been an important issue in the development of videotex worldwide. PresTel is the standard in the U.K. and is the basis for the standard in many parts of Europe, while NAPLPS (North American Presentation Level Protocol System) is becoming more popular in the U.S. because of its enhanced graphics capabilities. In the past, the unique graphics designs of Japanese videotex have precluded standardization with other international systems. However, with advances in technology and increased competition, the NAPLPS protocol has been making serious inroads into the Japanese videotex industry.

Recently, there has been a flurry of videotex activity in Japan that has been taking a number of forms. In May 1985, Nippon Telephone and Telegraph (NTT) announced the development of applications software that can directly convert NAPLPS frames into the CAPTAIN protocol. Earlier in 1985, NTT and C. Itoh agreed to target the business corporations and local governments with their own specialized CAPTAIN videotex service.

American Telephone and Telegraph Company (AT&T) is teaming up with Videotex Japan Network (VJN) to initiate a videotex service in upwards of 40 Japanese cities. They plan to use NAPLPS frames and, in the initial stages, install public terminals in Japanese hotels for providing information in English. In September 1984, Mitsui and Company announced plans to begin a nationwide videotex service in Japan on a franchise basis. They intend to install 3800 Canadian Telidon terminals, which also use the NAPLPS standard. NAPLPS will be the format for two more videotex joint ventures, one involving Mitsui and Company and the Sony Corporation and one that will include Mitsubishi, Secom, and Kyocera.

Since the privatization of NTT in April of 1985, competition in the Japanese videotex industry has increased dramatically. Seeking high volume markets, the competitors to the NTT CAPTAIN system have been offering products that are compatible with the increasingly popular NAPLPS standard. Despite its national goals, it is clear that Japan cannot be a leader in all aspects of every new information industry endeavor.

If Japan is not a captain on the videotex vessel as we know it today, it may rise within the ranks in the future. Videotex is still evolving and will continue to be redefined in the years ahead. The technology is unproven, and the seas uncharted. On the horizon is the integration of videotex systems with artificial intelligence. We may soon see systems that anticipate the interests and needs of the user. Intelligent videotex products may be able to present tailored information from their databases, eliminating the need for tedious keyword searches and menu browsing. Japan's Fifth Generation project is expanding and sharpening the country's expertise in artificial intelligence. Perhaps they will become leaders in the design and development of a new generation of intelligent videotex systems.

References and Further Readings

Doe, P. "INS: Bullet Train of Telecommunications." *Electronic Business,* 10(13)1984:102–105.

Feigenbaum, E.A. and P. McCorduck. *The Fifth Generation.* Reading, Mass.: Addison-Wesley, 1983, p. 15.

Hopper R. "Lessons from Overseas: The British Experience." In Martin Greenberger, ed., *Electronic Publishing Plus.* New York: Knowledge Industry Publications, 1985.

Sigel E. *The Future of Videotext.* New York: Knowledge Industry Publications, 1983, p. 156.

Tydeman, J., et al. *Teletext and Videotext in the United States.* New York: McGraw-Hill, 1982, p. 2.

REPORT

The Future of New Media in the Home

Andrew P. Hardy
General Electric Corporation

The early 1980s have seen a number of new products and services being offered for home entertainment. Chief among them have been videogames, home computers, video-cassette recorders, cable television, and video-disk players. Each of these technologies has had its own unique history. Some have had great but short-lived success. Others occupy an increasing amount of our time at home.

Videogames were the first of these products to gain widespread interest. They were quickly followed by the home computer and video-cassette recorder. Cable television, first available in the 1960s, has now become another important source of home entertainment in the U.S. These communication innovations have not arrived on the scene unnoticed. Videogames fast became part of our popular culture, then disappeared almost as quickly as they came. The home computer appears to have taken the place of videogames in the public mind. Video technologies, like cable television and video-cassette recorders, have also generated a great deal of interest.

Unfortunately, we often get caught up in the excitement of the moment when thinking about the possibilities of these new media. We worry about videogames corrupting the youth of the nation. Parents feel that their children will flunk out of college if they don't have the benefit of a computer education. The television industry wonders if people will seek video alternatives to their programming. The excitement of the moment can also create false impressions about how widespread the use of these new forms of home entertainment is and how many people may ultimately purchase them.

Not every new offering in the home entertainment field will be purchased by everyone. Complete acceptance of a new medium takes some time to occur. For example, during the height of the popularity of videogames, about 50 percent of the households in the U.S. had no interest in purchasing a videogame unit. No more than 35 percent of the homes in the U.S. had actually purchased a videogame unit. Even the TV set took at least ten years to become a universal medium of home entertainment.

This raises two questions. Which of the new forms of home entertainment will become widespread? How quickly will they become so?

As a start to answering these questions we can consider the various factors that influence the purchase of these new media. The first set of factors to consider is the kind of entertainment these technologies offer and the desires of the public. Does the public really have a need or desire that can be satisfied by these media? The second set of factors is how much do they cost and how much is the public willing to pay?

The home computer can be used as an example of this viewpoint. We can first discuss consumer needs. The major functions for home computers today are game playing, cataloging information, and word processing. These are also the major kinds of software that are produced today. Not everyone has a need for writing papers or keeping a library card catalog at home. Surveys indicate that the main reason for people not to buy a home computer is that they have no need for one.

Many people are not willing to pay thousands of dollars for a home computer. Surveys indicate that high cost is the second major reason for not purchasing a home computer.

With these factors in mind, the ability to satisfy an entertainment need and the cost, let us venture some guesses about how home entertainment media will prosper in the future.

The future for these media does appear bright. Whether incorporated into a home entertainment center or as a hodgepodge of devices, new home entertainment media do promise to become more widespread. Surveys indicate that by the end of the 1980s, some form of these new home entertainment media may be found in as many as 65 percent of the homes in the U.S. Optimistic estimates would say that as many as 60 percent of the homes in the U.S. may have a number of these new media, such as home computers, video-cassette recorders, and cable television.

As has been found with many other innovations, from agricultural products like hybrid seed corn to consumer durable goods like washing machines, a key attribute of these households is family income. Higher income families always seem to be the leading purchasers of new products. New home entertainment media do not seem to be different. These products represent an expensive purchase for most people. Those families with great financial resources are more likely to purchase a new product like a home computer, because they can afford it. In addition, higher income families will be able to afford many more of these new media.

The key factors that will control the extent and pace of growth of these new media will be their cost and the ability to meet the needs of the consumers. The price of these new media continues to decline, while their capabilities continue to increase. The trend in the computer industry tends to be to build computers that have more memory and greater speed at lower cost. We also witness the same phenomenon in the area of video technology. Each day new products are developed with more features and lower cost. This factor appears to be favorable for continued growth of these new media.

It is the ability of these new media to deliver what the consumer wants that will probably control their growth. What these new media appear to lack is adequate software in the case of computers, and programming in the case of video media. New home computers are continually introduced to consumers. What they universally lack at first introduction is software to perform the things that consumers want. Cable systems improve their offerings by adding new channels. Unfortunately, there doesn't seem to be enough quality programming to fill all the time and channels available to the viewer. One medium has an advantage in this regard. Video-cassette recorders are used for time shifting and can also let viewers skim off the cream of programming for their own use. In addition, movie studios and television networks have stockpiles of old movies and programs that can be made available for consumers. This advantage would appear to give video technology, especially VCRs, the greatest potential for growth in the near future.

References and Further Readings

Dickerson, M.D. and J.W. Gentry. "Characteristics of Adopters and Nonadopters of Home Computers." *Journal of Consumer Research* 10(1983):225–235.

Gatignon, H. and T.S. Robertson. "A Propositional Inventory for New Diffusion Research." *Journal of Consumer Research* 11(1985):849–867.

Rogers, E.M. *Diffusion of Innovations*. New York: Free Press, 1983.

Topics for Research or Discussion

▬▬▬▬ One long-held generalization in the study of communication media is that one medium usually does not replace another. Instead, when new media are adopted, uses of the old media change. For example, it is said that television largely replaced the B movie. Yet the movie business still exists.

What observations might you make about the effects of video cassettes on traditional broadcasting and theater business? What are the effects of audio disks upon other home entertainment? What might be the effect of cable television services upon traditional broadcast television services? Select one example and examine it in detail.

■■■■■■ Do you think that video display terminals change the way you write or edit? What are your own experiences with word processors? As software for such equipment gains efficient spelling checkers, thesaurus lists, and grammatical evaluators, will writers and reporters forget the rules? Or will such devices allow them to concentrate upon meaning and style?

■■■■■■ Analyze a shopping center in terms of its information content (e.g., getting the attention of for the consumer, reducing search time, presenting optimal mixes of services, advantages to the retailer in terms of shopper captivity, and low price competition within the mall).

■■■■■■ Several popular computer communication network services are available to the public (e.g., The Source and CompuServe). Prepare a brief report on what these services offer, including a general picture of the costs involved. Describe the types of equipment that you need to use the service and the steps involved in obtaining a subscription. Include in your report a description of some of the specific uses you might make of such a service.

■■■■■■ Consider some of the effects that the changing environment of public communication is having upon advertising. For example, some analysts say that special forms of cable broadcasting, particularly text services, will never succeed simply on the basis of subscription fees; they will have to be attractive to advertising. Do some research on this topic and interpret what you consider to be some of the specialized opportunities for new forms of advertising. What are some of the consequences for the structure of the advertising message?

References and Further Readings

Arlen, G. "The End of the Beginning: What Will We Do with What We Learned on the Trial Trail?" In *Videotex '83 Proceedings*. New York: London Online, 1983, 337–343.

Barnouw, E. *Tube of Plenty: The Evolution of American Television*. Oxford: University Press, 1982.

Bernstein, P.W. "Television's Expanding World." *Fortune* 99(13):64–9 (1979).

Broadcasting. "Special Report: SMATV—The Medium That's Making Cable Nervous," (June 21, 1982) 33–46.

Broadcasting. "Special Report: Fine Tuning the Technological Explosion," (December 17, 1979) 35–42.

Broadcasting. "Teletext and Videotext: Jockeying for Position in the Information Age." (June 28, 1982) 37–49.

Compaine, B.M. *The Newspaper Industry in the 1980s: An Assessment of Economics and Technology.* White Plains, N.Y.: Knowledge Industry Publications, 1980.

Comstock, G., S. Chaffee, N. Katzman, M. McCombs, and D. Roberts. *Television and Human Behavior.* New York: Columbia University Press, 1978.

Dimmick, J. and E. Rothenbuhler. "The Theory of the Niche: Measuring Competition Between the Cable and Broadcast Industries." Paper presented at the International Communication Association Convention, Dallas, 1983.

Editor and Publisher. "Adapting to Satellites: Christian Science Monitor Says It Is Committed to Converting Its Plants around the Country to Satellite Technology." 117:38, Oct. 13, 1984.

Editor and Publisher. "AP Finds Meager Demand for Electronic News." 10, Oct. 2, 1982.

Editor and Publisher. "Electronic Newspaper Found Unprofitable." Aug. 28, 1982.

Editor and Publisher. "Viewtron Users Rate News as Top Choice." July 10, 15, 1982.

Editor and Publisher. "Wall Street Journal Completes Satellite Network." 115:20, Dec. 18, 1982.

Journal of Communication. Special Issue on Cable Television. 28(2): (1978).

McCavitt, W. E. *Television Technology: Alternative Communication Systems.* Layman, Md.: University Press of America, 1983.

McLuhan, M. *The Gutenberg Galaxy.* Toronto: University of Toronto Press, 1962.

Randall, S.D. "Effect of Electronic Editing on Error Rate of Newspaper." *Journalism Quarterly* 56 (Spring 1979):161–165.

Russell, N. "The Impact of Facsimile Transmission." *Journalism Quarterly* 58 (Autumn 1981): 406–410.

Shipley, L.J. and J.K. Gentry. "How Electronic Editing Equipment Affects Editing Performance." *Journalism Quarterly* 58(Autumn 1981): 371–374.

Media Businesses and Social Applications

Another view of public communication in free enterprise economies is the contrast between media as profit-making businesses and as public services. Education is likely our greatest investment in communication as a public service, and it has been given an entire chapter (11). In the present chapter, we examine several aspects of the media business and then turn to examples of public service applications.

TOPICAL OUTLINE

Communication as a Business
Media and Information Industries
Advertising as a Major Revenue Source
Cable as an Example of New Technology Growth
Can Growth Be Predicted?

Are Tastes Changing?

Social Applications of Communication
The "Sociologizing" Mode
Transportation Tradeoffs

COMMUNICATION AS A BUSINESS

Media and Information Industries

Too often in the study of modern communication we academics fail to give sufficient stress to the fact that most media or communications organizations in the Western world are businesses to be operated for profit. Ultimately, then, the criterion of profitability rather than the quality or nature of the medium itself may be the standard by which a particular medium stands or falls. Let us look first at some statistics on information businesses, as summarized in Table 10.1.

A first generalization from these statistics is that far more revenue is generated by so-called information technology businesses (e.g., IBM and AT&T) than by traditional mass media organizations. Another, although not shown in these data, is that growth figures favor the former over the latter. And, as mentioned earlier in this volume, much of the growth of information technology is within investment in reorganizing traditional industries.

But there is another important consideration: Many mass media businesses depend upon advertising rather than unit sales as their major source of revenue. In this respect, they are a different form of business: They are distributors of advertising messages rather than a newspaper, magazine, or broadcast program as a product.

Advertising as a Major Revenue Source

The key basis for setting value on a medium's ability to generate advertising revenue is the size and sometimes the quality of the audience it can reach. Probably the most publicly visible examples of audience evaluation are television rating services. For example, the Nielsen Company selects a nationwide sample of 1500 households that are taken as representative of the U.S. television-viewing audience, and the Audit Bureau of Circulation presents validated figures of the circulation of major periodicals. Table 10.2 summarizes some of the gross revenue figures for the advertising business in America.

It is usually a surprise to people that the direct mail business in America reports substantially more advertising revenues (about 37 percent of the total in 1984, omitting the "other" category from the total) than broadcast or publications businesses. Broadcasting, especially television, accounts for approximately 40 percent of the total, as against around 20 percent for newspapers and magazines combined. You might also note in Table 10.2 that direct mail has a slightly greater growth rate.

Cable as an Example of New Technology Growth

A key question about new communication technologies, especially those that might directly compete with traditional media, is the degree to which they can capture advertising revenues. Although cable began with, and continues to be,

Table 10.1 *Revenues of Selected U.S. Communication Businesses*

Based on data for 1985 provided by Media General Financial Services, Richmond, Va. All amounts are given in millions.

Name	Annual Revenue	Earnings
Magazines:		
Meredith	533.4	42.7
Playboy	192.3	4.2
Time Inc	3403.6	199.8
Cardiff Comm	5.7	−1.3
Frost Sullivan	13.8	0.4
Webb	177.3	7.0
Book Publishers:		
Harcourt	818.9	50.5
Harper Row	201.4	7.7
Houghton Miff	277.5	18.9
Macmillan Inc.	676.9	44.7
McGraw Hill	1491.2	147.4
Addison Wesley	139.4	6.7
Scholastic Inc	180.3	3.0
Waverly Press	65.2	2.6
Wiley John	213.4	7.6
Newspapers:		
Belo AH	385.2	23.8
Dow Jones	1039.3	138.6
Gannett Co	2209.4	253.3
Knight Ridder	1729.6	132.7
Times Mirror	2946.7	237.1
Tribune Co	1937.9	123.8
Media General	578.6	32.8
NY Times	1393.8	116.3
Wash Post	1078.7	114.3

Table 10.1 *(continued)*

Based on data for 1985 provided by Media General Financial Services, Richmond, Va. All amounts are given in millions.

Name	Annual Revenue	Earnings
Broadcasting		
Cap Cities ABC	1020.9	142.2
CBS Inc	4676.8	27.4
Chriscraft	190.3	18.3
Rollins Comm	112.4	13.2
Taft Brdcst	472.8	19.4
Price Comm	41.8	−12.3
Turner Brdcst	351.9	17.3
Telecommunications		
Am Tel T	34909.5	1556.8
Ameritech	9021.1	1077.7
Bell Atlantic	9084.2	1092.9
Bell Canada	13257.4	1050.8
Bellsouth	9175.1	1417.8
Comsat	459.0	−41.5
GTE	15732.4	−161.1
NYNEX	10313.6	1095.3
Pac Telesis	8498.6	929.1
So N Eng Tel	1304.0	119.9
Swtrn Bell	7925.0	996.2
US West	7812.6	925.6
Wstn Union	1082.5	−371.4
MCI Comm	2542.3	139.6
Computers		
Burroughs	5037.7	248.2
Centronics Data	216.3	0.2
Compaq	503.9	26.6
Control Data	3679.7	−567.5

(continued)

Table 10.1 *(continued)*

Based on data for 1985 provided by Media General Financial Services, Richmond, Va. All amounts are given in millions.

Name	Annual Revenue	Earnings
Cray Research	380.2	75.6
Data General	1239.0	24.3
Datapoint	520.2	–51.2
Digital Equip	7590.4	617.4
Hewlett Pack	6505.0	489.0
Honeywell Inc	6624.6	275.4
IBM	50056.0	6555.0

mostly a subscription-fee medium, many analysts see its eventual major growth in the form of advertising revenues. Tables 10.3 and 10.4 illustrate some projections.

In both these tables, cable is seen as a major competitor for advertising dollars. The question is which medium will lose these dollars—broadcasting or print.

Can Growth Be Predicted?

If you read the newspapers, you cannot escape the fact that communications and computing businesses are often volatile areas of growth. As with all

Table 10.2 *Advertising Revenues (in millions of dollars)*

Medium	1982	1983
Network TV	$6210	6985
Spot TV	4360	4820
Radio	1178	1320
Magazines	3710	4210
Newspapers	2452	2685
Direct Mail	10319	11765
Other	9556	10450
Total	37785	42235

From "Ad pages to increase 8% to 10% in 1984, Coen predicts." *Folio* 13(2): 16ff, February.

Table 10.3 *Cable Network Advertising Revenue (in millions of dollars)*

	1983	1984	1985
WTBS	$134	158	180
MTV	25	52	83
ESPN	42	58	72
USA	30	34	50
CNN	40	52	64
CBN	13	26	39
OTHERS	21.7	27.5	82.5
Total	305.7	437.5	566.6

From CableVision 10(2): 100, Sept. 10, 1984.

businesses, they are affected by fluctuations in the economy, changing needs of consumers, and international competition. But several clusters of unique factors also affect these businesses. One is that directly or indirectly telecommunications and to some extent computing have been influenced by governmental regulation. That is, quite apart from the usual business climate, many aspects of these businesses have grown as a result of governmental influence (e.g., the former AT&T monopoly) or have been inhibited by barriers to certain areas of growth (e.g., IBM's long antitrust battles). Although most such influence is the consequence of legislative processes, outcomes are not always predictable. For example, most experts did not expect

Table 10.4 *Projected Growth of Cable Advertising Revenues (in billions of dollars)*

1983	0.353
1984	0.514
1985	0.725
1986	0.995
1987	1.353
1988	1.778
1989	2.238
1990	2.765
1991	3.349
1992	4.045
1993	4.872
1994	5.714

From Marketing and Media Decisions, 19(6): 26, May 1984; Paul Kagan and Associates.

the breakup of the Bell system to come so abruptly nor with many of the conditions that were imposed.

Another cluster of factors affecting growth is the rapid change in the technologies themselves. Products, especially in computing, often have short "product lives" before new advances render them obsolete. It is not unusual for sales of a product or service to fall prematurely flat when rumors abound that a new model is about to appear or that the competition is soon to offer a superior product.

Given this volatility, growth and change in the communications business have often been accused of being more by chance or whim than by conscious planning. In the first of this chapter's reports, William Dutton advances a fresh perspective, suggesting that we examine growth from the logical context of games. Growth may not be so irrational as it is the culmination of interacting interests, including the law, the economy, and technological change.

ARE TASTES CHANGING?

It is not evident whether the growth of new media is one-for-one at the expense of traditional media. True, there are industry analyses now and then that describe replacement—for example, the dip in record sales when video games were popular, or the drop in network television viewing with the rise in cable usage. Yet, it appears more as if the overall use of media is increasing; one medium is not so much replacing another, as is so often generalized in the study of communications. There are changes in the patterns of usage. To use a transportation analogy, we do not have such striking examples of media replacement as, for example, the growth of air travel coupled with the demise of passenger trains.

Another way to view possible change is to study the habits or attitudes of the younger population, assuming that as they become the mainstream of the media marketplace, they will bring their new tastes with them. However, despite the tastes of, say, university students in music or movie-going, the author has usually found their anticipated gratifications from media to be similar to the mainstream. For example, it has been popularly advanced that television replaces newspapers as the chief informational medium for university students, yet most data are to the contrary.

SOCIAL APPLICATIONS OF COMMUNICATION

The "Sociologizing" Mode

In his forecasts of the nature of postindustrial society, Daniel Bell (1976) also introduces his concept of the "sociologizing" mode in socioeconomic growth.

This is investment in services that are aimed at the public good, which are by their basic nature not profit-making enterprises but social investments.

This is not a new concept in public communication, for in most countries of the world many mass media are either operated or subsidized by government as public institutions. Postal and telecommunication services are another major example; the U.S. contrasts with most of the world in having commercially based telephone services with a trend toward increased deregulation.

In the remainder of this chapter, we will examine social applications of public communication. One is the substitution of communication for transportation services, a general concept applicable to a wide variety implementations, such as education, health, and work environments. Finally, in reports by William Paisley and Martin Elton, we will see concrete examples of sociologizing applications of public communication.

Transportation Tradeoffs

Whether it is called *telecommunication-transportation tradeoff, telecommuting,* or *teleworking,* the telephone and other network installations offer numerous opportunities for substituting communication for transportation in modern society. Perhaps the most public version of this for the telephone is the Yellow Pages advertising slogan, "Let your fingers do the walking."

Trading communication for transportation has become a topic of special interest to city planners as they calculate the problems of overcrowding, pollution, streets, and parking that accompany a large influx of workers to a city's businesses and industries. If it were possible to disperse the industries or their offices to outlying areas and then serve their management needs by a telecommunications network, the travel time of many workers would be shortened. An extension of this idea is using the telephone network to make certain types of work available in the home. This is particularly applicable to some types of information work, such as answering correspondence, creating documents, computer programming, or monitoring or interacting with financial services.

Many arguments have been advanced that the savings in energy alone would justify use of the telephone network for communication-transportation tradeoffs. For example, in a study sponsored by the National Science Foundation, Jack Nilles and his colleagues (1976) determined that the energy consumed by private automobiles was about 30 times the energy needed by the telecommunications that might substitute for the work accomplished. The researchers concluded that a mere one-percent reduction in commuting could save enough fuel to provide electrical power to a medium-sized city.

Despite all the arguments advanced in favor of using the telephone network for communication-transportation tradeoffs, there is no evidence of a major growth in this direction. As might be expected, some of the frustration in decentralizing the workplace is that it places strains on the managerial ability to supervise and otherwise coordinate a workflow. Although the telecommunications links may allow the work to be transferred to the worker, managerial processes, at least under the present practices, are not so easily stretched. There is also the cost of remote work sites, including the practical problem of a necessary dedicated telephone line into a worker's home. The costs of telephone installation, dedicated line services, and whatever additional information technologies are needed in the home, all add up to a sizeable investment for the home worker.

Another factor that has emerged in working at home is that many individuals simply do not want to stay in the home for a day. They prefer the change of going to the office. Or, put another way, they do not like to disrupt the more leisurely atmosphere of their home with the pressures associated with a workspace. One further observation, particularly by labor unions, is that companies who encourage part-time home work, often tend to do so at the expense of full-time unionized workers. Also, it is asserted that minimum or below-minimum wages are often paid to home-station clerical personnel.

In brief, although the wired network has become one of the newest areas of interest of urban planners for the above reasons, many of the practical challenges of decentralized, or home, work may prevent the rapid adoptions of telecommunication and transportation tradeoffs.

REPORT

Technology Growth as an Ecology of Games

William H. Dutton
University of Southern California

Telecommunication systems are undergoing change with the introduction of new technologies and services. While a great deal is known about the technology that is driving this change and much about the various developments that have been proposed, there is relatively little understanding of the process by which decisions about these developments are being made. From an *ecology of games* perspective, first elaborated by Norton Long (1958) as a description of land-use decision making in metropolitan areas, the development of telecommunications for society as a whole evolves from the interactions of individual games, each with its own rules, actors, goals, and strategies. The different games are interrelated by some common rules, shared beliefs, and political-administrative traditions, and by some actors that simultaneously participate in different games or transfer from one game to another. The outcome is the net result of the individual games and their interactions—that is, an ecology of games.

In this respect, no one governs the development of communications in the rational-comprehensive sense that the act of governing might connote. Instead, actors make decisions about real estate investments, cable franchises, and so forth that affect the pace and direction of the development of telecommunications. The value of this perspective has been suggested by a variety of case studies. One of these is presented below to illustrate this perspective and help provide a more concrete sense of its application.

The Development of a Microwave Communications Network

From an ecology of games perspective, the development of telecommunications infrastructures and services is a product of the interaction of different games being played by different players, each with somewhat different likes and interests. Therefore, the nature of developments not only is difficult to predict, veering from any trajectory formed by extrapolating from the history of individual players, but also is likely to differ from the strategic plans of the players. The movement of a large publishing company into the microwave communications business provides an illustrative case of such a process.

The Times Mirror Company

The Times Mirror Company is a $2.8 billion a year media communications company (The Times Mirror Company, 1984; *Los Angeles Times,* 1985, p. 16). Headquartered in Los Angeles, California, the company and its subsidiaries are located in over two dozen cities across the United States. Its principal revenue sources are newspaper publishing, newsprint and forest products, book publishing, information services, including flight information and training programs, charts, maps, and directories, cable television, including videotex and microwave communications as well as cable, and broadcast television, in addition to other operations, including magazine publishing and art and graphics products. The print media— newspaper, book, magazine, and other publishing—account for nearly 80 percent of the company's revenues. The remaining 20 percent involve the newer electronic media, including cable, videotex, software publishing, and broadcasting.

The company was incorporated in 1884 for $40,000, joining the *Los Angeles Times* (initially called the *Daily Times*) with the Mirror Printing Office and Book Bindery, which is now Times Mirror Press (Berges, 1984). With the success of the *Los Angeles Times* newspaper, the company acquired newsprint mills and forest lands in Oregon and Washington in 1948 to control its supply of newsprint. The company began to diversify in the 1960s and 1970s by acquiring businesses in the United States and abroad that were known for their quality. By 1983, the company was still primarily in print publishing, but its holdings placed it squarely in the newer electronic media as well. In the 1980s, the company more explicitly defined its business to be in the information and communication industry, leading the company to sell some businesses that were tangential to its focus. In addition, it acquired other media, communications, and information-related businesses and began efforts to concentrate its cable holdings within regions by trading franchises (subscribers) with other MSOs with similar business strategies. Another feature of the 1980s was a decision by the company to launch internal development efforts aimed at the creation of new services and businesses, which fueled the company's move into the videotex and software publishing business in the 1980s (Baer, 1985). Throughout this history, it nurtured a traditional operating philosophy of granting a great deal of autonomy to its individual operating units—a tradition that is linked to the company's early history of guarding the editorial independence of the *Los Angeles Times*.

Cable Television

Following World War II, in 1948, the same year that the company acquired its newspaper mills, Times Mirror acquired its first cable company, a small system near its headquarters which served parts of Long Beach, California. But this initial move into cable television was an exception for the company until the 1960s when the company began to enter into electronic communications and information services, with broadcast and cable television being important areas of activity. Times Mirror moved into broadcasting through the purchase of television stations in Dallas and Austin, Texas, and five additional television stations from the Newhouse Newspaper Company. In 1979, the company purchased additional cable operations, acquiring Communications Properties, Inc. By 1984, the company's cable subsidiary, Times Mirror Cable Television, Inc. (MCT), rose from owning a single local franchise in the Los Angeles region to being one of the seven largest multiple system operators (MSOs) in the United States, operating 61 cable systems serving nearly one million subscribers in over 300 communities in 16 states (Times Mirror Company, 1984).

Times Mirror Microwave Communications Company

Cable originated as a community antenna television system, amplifying and retransmitting over-the-air signals received by a master antenna to the community's cable subscribers. In 1965, through the Federal Communication Commission's First Order and Report, operators were restricted from duplication (retransmitting a program from a distant station that was being shown by the local broadcast station). Then, broader restrictions on the importation of *distant signals* (those originating from points too distant to be received by ordinary television receivers) were established by the FCC's Second Order and Report, in 1966, and then again in 1968. These restrictions undercut the ability of cable to offer more choice in programming, a principal selling point in urban areas, but it was felt to protect the markets of local broadcasters. Since that time, however, the courts and the FCC have lifted most restrictions on the importation of distant signals, and reception of these signals has become an attractive feature offered to cable subscribers, particularly in small towns and cities that would otherwise not have access to the independent television stations of the larger cities of their region. In most cases, this was accomplished by installing a microwave communications link between the television station and the head end of the cable networks.

In 1979, Times Mirror acquired Communications Properties, Inc. (CPI), a cable television company that served communities in parts of Texas and Ohio. In acquiring the cable television facilities, it acquired three small microwave companies in Texas and Ohio as well. The microwave facilities were used primarily to import video signals to CPI's cable companies, but also to distribute programming to network affiliates. However, by 1979 the utility of microwave networks by cable operators and network affiliates had begun to decline with the development of satellite communications to link network television stations and cable systems. Compared to satellite communications, a microwave communications network for cable systems, which requires only one-way distribution of the video signal to more than one location, was less economical. Thus, microwave communications for the distribution of television programming to television stations and cable systems was thought by many to be a declining business (Baer, 1985).

Given these bleak prospects for microwave distribution of television signals, the company considered selling its microwave facilities soon after CPI was acquired. However, the general manager of the Microwave Division (now Times Mirror Microwave Communications Company, or TMMCC) saw an opportunity to lease some of the microwave facilities to a young company in the voice and data business, MCI Communications. MCI was interested in leasing circuits in Texas because it did not have its own network fully in place and did not wish to obtain circuits from its competitor, AT&T, if there were other options. A similar deal to lease microwave circuits in Ohio soon was struck with Southern Pacific-Sprint (now GTE Sprint), another emerging competitor of AT&T in the long-distance business (Baer, 1985).

These initial moves into a new business area were undertaken by the general manager of TMMCC on his own initiative, without any strategic direction from Times Mirror corporate. This action was consistent with Times Mirror's general policy of decentralizing authority and responsibility. It was not until the TMMCC general manager requested major new funding to upgrade and expand the microwave network that corporate management became involved. At that time, a detailed study of the U.S. long-distance business was undertaken, alternatives were evaluated, and a business plan was prepared that would position TMMCC as a "carrier's carrier." Based on this study, as well as on TMMCC's early success in repositioning its facilities to serve a new class of customers, Times Mirror became convinced that a window of opportunity existed for wholesaling microwave circuits to a growing group of entrants into the long-distance voice and data communications business (Baer, 1985).

The argument in support of this venture was both technological and financial. Technologically, advancements in microwave relays had made them increasingly well suited to voice and data communication. And the technology was already well developed and easily put into place, compared to other terrestrial communications systems. While fiber optics seemed to define the future of long-distance communication, a microwave network could be constructed years before a fiber-optics system could be put in place. And for point-to-point voice and data communication, in contrast to one-way broadcasting of video signals, satellites were not yet developed to a point that they were as well suited. Financially, the company had the resources to construct such a network, while the emerging long-distance companies did not yet have the traffic or the capacity and resources to construct their own networks in all parts of the nation. For a time, the Times Mirror Microwave Communications Company could provide the new long-distance companies with an economical alternative to leasing from AT&T.

By 1985, the subsidiary had upgraded systems and added nodes on the network to the point that it had installed a terrestrial microwave communications network that spanned the nation, from Los Angeles to Pittsburgh, with plans for its extension to other east coast cities. In 1984, the company expected the network to reach 31 cities in 17 states upon its completion (Times Mirror Company, 1984). Serving such common carriers as MCI, Sprint, and ITT, Times Mirror Microwave Communications Company has become an important actor in voice and data communications—a business that few would have expected Times Mirror to enter based only on its strategic plans and past acquisitions.

The Decision-Making Process

In the late 1970s, the company's business plans did not envision a major move into the microwave communications business. To the Times Mirror Company, this was a rational and successful response to a target of opportunity (Baer, 1985). And in hindsight, there was an environment at Times Mirror that fostered such ventures in that it had (since the days of its early founder, Harrison Gray Otis and his son-in-law, Harry Chandler) a tradition of valuing entrepreneurial executives; a management philosophy that permitted its operating companies to make important decisions; an emphasis on the internal development of new businesses; and a definition of the company that could incorporate a venture into a new communications business.

The development and design of this coast-to-coast terrestrial communications network was not planned long in advance by a small group in corporate headquarters, nor was it the product of a pluralistic process of bargaining and negotiation among all the major actors. It is unlikely that any actor involved in this business made the decision or wrote the plan to build the long-distance network. Actors made decisions about other things, such as whether to purchase a cable system and whether to sell or lease an under-utilized microwave facility. The network evolved through the interaction of these incremental business and regulatory decisions made by the Times Mirror Company, its subsidiaries, several common carriers, and state and federal regulatory agencies.

References and Further Readings

Baer, W.S. "Electronic Publishing at the Times Mirror." Presentation at the Annenberg School of Communications, April 9, 1985, followed by interviews with the first author.

Berges, M. *The Life and Times of Los Angeles*. New York: Atheneum, 1984.

Dutton, W.H. with Helena Makinen. "Urban Telecommunications as an Ecology of Games." Paper presented IRIS '85 and Municipal Information Systems, Pacific Area Community (MISPAC) Seminar in Tokyo, Japan, July 3–14, 1985.

Long, N.E. "The Local Community as an Ecology of Games." *The American Journal of Sociology*, 64(1985):251–261.

Los Angeles Times. "California Leading Companies 1985." June 9, 1985.

Times Mirror Corporation. *Times Mirror Factbook 1984*. Los Angeles: Times Mirror Company, 1984.

REPORT

Lessons of a Videotext Trial

William Paisley
Edupro

Teletext, videotext, microcomputers, interactive video disks, and other microprocessor-based communication technologies have a strong family resemblance. They can all retrieve information on demand and display it in the form of text or graphics. However, each of these new media can perform some communication functions better than others. In the decade of the 1980s, many experiments are being conducted to find the best match between new media and new functions.

Of special interest are the tradeoffs between teletext and videotext, which will compete with older media such as newspapers and television for the future information audience. Teletext and videotext can be indistinguishable on the screen, but they are based on different transmission technologies and have different capabilities and costs.

Teletext refers to electronic text transmitted during the vertical blanking interval of a broadcast television signal, although it can also be transmitted in other channels such as FM subcarriers. Teletext frames are broadcast in a constantly repeated sequence. The number of frames is limited by the length of time the sequence takes to repeat, since the user must wait for a given frame to be broadcast before it can be captured by the teletext decoder and displayed on the screen. Transmitting five frames per second (highly detailed, or high-resolution, frames take longer), a teletext system can offer about a thousand frames of information at a time. To minimize waiting time for popular frames, these are broadcast two or more times in each sequence.

The content of a typical teletext system, such as England's Ceefax/Oracle, reflects the premium placed on capacity. News, sports, weather, and financial information, timetables and event calendars, directories, advertisements, and shopping catalogs are the favored content of a teletext system because they are concise, widely used, and in some cases profitable.

Videotext refers to text transmitted via telephone or cable. Unlike teletext, which merely captures a signal that is being broadcast anyway, videotext requires a request channel from the user to the distribution center. Requested frames are sent to the individual user alone. Because it is not necessary to broadcast videotext frames in a frequently repeated sequence, the content of a videotext system can be orders of magnitude larger than the content of a teletext system. For example, one early videotext system, Channel 2000 in Columbus, Ohio, held a 32,000-page encyclopedia and the card catalog of a large public library as well as other extensive resources.

It is evident that teletext, which has a free ride on television signals, should cost the user less in monthly charges than videotext, which requires at least a brief two-way interconnection for each user. Videotext can become prohibitively expensive if a two-way interconnection is maintained throughout a long session (think of this in terms of telephone charges plus computer time). Therefore, a hybrid form of videotext called *dump-and-disconnect* uses the interconnection only long enough to receive a request and transmit the requested frames. Once the decoder has stored the requested frames in its memory, it disconnects itself from the line. The user can read these frames and hold them for later reference without incurring charges.

A dump-and-disconnect videotext system for farmers was launched by the U.S. Department of Agriculture in 1980. Like many other businesses, farming has become information-based. A farmer's margin between profit and loss is narrow; timely information on market futures, weather forecasts, soil conditions, fertilizer and pesticide costs, and even political developments (such as production support levels and grain embargoes) helps the farmer to make the profitable decisions. USDA's Green Thumb system was first tested in 200 farm homes in Kentucky. In addition to market, weather, and farm production advisories, Green Thumb contained information on home management, 4-H/youth, and community events. The total number of frames in the system fluctuated from month to month but averaged 250 in a typical month. Major categories of information such as commodity prices, weather forecasts, and agronomy were covered in about 30 frames each.

Green Thumb was indeed a videotext "trial" to the Kentucky farmers. Sometimes the system worked; sometimes it flashed "abort"; sometimes it didn't work at all. Because of the dump-and-disconnect feature, farmers had to enter their full requests before learning whether the system was "up" or "down." At the midpoint of the trial, weather information failed to update about 60 percent of the time (the record subsequently improved to about 20 percent of the time). Late in the trial, because of a computer change hundreds of miles away, commodity price updates ceased for seven weeks.

Despite the logistical mishaps that can be expected in a trial of this kind, Green Thumb provided valuable information to the Kentucky farmers. According to an evaluation conducted by researchers at the University of Kentucky and Stanford University, the 200 farm families called Green Thumb 28,399 times in the course of a year, for an average of 142 calls per family. Many stories were told of farming mistakes avoided and money saved. (For example, one farmer estimated saving $7000 through a more limited use of pesticide, thanks to the county-by-county pest advisories).

The success of the Green Thumb trial has already led to a number of other agricultural videotext systems operated by public agencies and private companies. However, the lessons taught by Green Thumb have farther-reaching implications. The first lesson is that a wide-area videotext system that depends on a network of computers (e.g., for market and weather updates, for primary storage, and for call handling) will fall victim to Murphy's law: Whatever can go wrong, will, and at the worst moment. The call-handling computers located in each county were intended to save farmers the cost of long-distance calls to the main computer in the state capital; they also added one more link to an over-extended chain. In one county the call-handling computer was housed in the damp basement of the courthouse; it often "called in sick" to the main computer.

The second lesson concerned the users' patience with dump-and-disconnect systems. At the normal rate of 30 characters per second, each frame required more than 15 seconds to transfer to the decoder's memory. Because of this delay, users began to limit their requests to the marketing and weather information they most wanted to see. Thus a general-purpose system was reshaped by circumstance into a special-purpose marketing and weather information system.

Third, the decision to use videotext rather than teletext technology limited the number of users because of the number of computer ports that could receive calls. In turn, the small number of users became a factor in frame updating. Agricultural extension personnel who were responsible for thousands of families outside the Green Thumb experiment could not devote as much effort to frame updates as they wished.

Note that a 250-frame database is well within the capabilities of a teletext system. Furthermore, the Green Thumb counties had good television reception from nearby cities. A telecast audience of 2000 or 10,000 families, rather than the 200 families of the videotext experiment, would have justified more effort in frame updating. More timely information, faster capture per frame, and greater system reliability would have increased use rates.

In addition to these technological lessons, Green Thumb gave notice that the knowledge gap was not going to disappear in the era of new media. Education, age, and innovativeness were related to Green Thumb use in a familiar pattern. Farm size was also related to Green Thumb use, even though Green Thumb was a free service during the trial.

For those who use information effectively, the new media will provide more extensive and timely information than before. Others will be increasingly disadvantaged by what they don't know.

Further Readings

Bretz, R. *Media for Interactive Communication*. Beverly Hills, Calif.: Sage, 1983.

Paisley, W. "Computerizing Information: Lessons of a Videotext Trial." *Journal of Communication* 33(1983):153–161.

REPORT

Television and Health Care Delivery

Martin Elton
New York University

Many telemedicine projects were initiated during the course of the last decade. Most were as conceptually primitive as they were technologically advanced. Results were generally disappointing. The project described here is one of the exceptions. While it was conceptually and methodologically sophisticated, it was technologically low-key, and it added considerably to understanding of the issues involved in using communication technology to improve access to health care in remote areas.

The program of research that led to the installation of the Northern Ontario Telemedicine System was triggered by two graduate students interested in extending U.S. research to Canada. The program was led by David Conrath, Ph.D. and Earl Dunn, M.D. who, along with their associates, have been most generous in discussing their findings with me over the years. Recently they have published a book about their work, and I have drawn heavily upon it in providing this case study.

In 1977 a telemedicine system was introduced into the Sioux Lookout Zone in northern Ontario. The region would seem to lend itself to telemedicine. It is a sparsely populated area, almost the size of California. Approximately 8000 Cree and Ojibway Indians live in 24 native villages, and another 2000 Indians live in the three predominantly nonnative towns. Communication is difficult. Only one of the villages can be reached by road. Most travel requires air transportation, which is expensive and sometimes impossible for days at a time. It is dangerous, too.

Since 1971 all communities have had high-frequency radio, but the atmospheric interference in the far north, coupled with equipment failures, have made it unreliable. Additionally, it is an open system and hence unsuitable for the exchange of confidential information about a patient. The more recent installation of telephone service in all the remote communities has been a considerable improvement. Access can be a problem, however, since some communities have only one or two lines. Another kind of problem is posed by close family ties. These make the transfer of a family member to the Zone Hospital a traumatic event.

Health care is provided by a regional system of the Canadian Department of National Health and Welfare. Facilities are a 70-bed hospital at Sioux Lookout and a network of nursing stations and health aide clinics in the villages. Physicians and specialists based at the Hospital make routine visits to the nursing stations and clinics. The nursing stations are staffed by two to five nurses and some ancillary personnel. Each station has an X-ray machine, laboratory facilities, and beds for several patients. The nurses also supervise the care provided at clinics in their satellite communities. The health aides who staff the clinics are local residents, selected jointly by the chief, the band council, and the Zone administration. Their training is a very basic six-week course, but many do not even go through this. The turnover of physicians and nurses is high: Average lengths of stay are less than two years.

In 1979, residents of the villages made an average of almost 5.5 visits per person to health care facilities. This was considerably more than for Ontario as a whole, where the average was 3.1.

Such was the environment into which David Conrath and Earl Dunn and their colleagues at the Department of Management Sciences, University of Waterloo, and the Faculty of Medicine, University of Toronto, respectively, introduced a telemedicine system based on slow-scan television. Its primary purpose was experimental, since it represented the final phase of a ten-year program of research. Earlier phases had involved detailed observation of the communication content of physician-patient consultations and controlled experiments to compare the effectiveness of different modes of telecommunication for diagnosis and treatment planning in primary care. The modes were color television, black and white television, slow scan television, and hands-free telephone.

The results of this research had been surprising. Provided a nurse was with the patient, there were no statistically significant differences among the modes in terms of the effectiveness and efficiency of diagnosis and patient management, nor between any of them and the conventional in-person consultation. (And a nurse would have to be present since touch was important in diagnosis, more so than visual cues). There were some differences among the modes relating to the physicians' confidence in their diagnoses and to the preferences of all concerned: In both respects, the more sensory-rich modes were superior but not overwhelmingly so. It appeared, then, that most of the telemedicine projects initiated in the previous few years were using unnecessarily expensive visual technology. (Cost is not the only issue. For example, the more sophisticated visual technology necessitates more training, requires more space, and raises more difficult repair problems.)

Although the research evidence for these conclusions was strong, only a field experiment could produce results that would be taken as definitive. For various reasons, results in the real world might differ from those in a laboratory setting. A visual communication system in the field would be used for purposes that had not been relevant in the contrived setting, for example, in particular for the transmission of X-rays.

The experiment involved the installation of a nine-station slow-scan television network. One unit was located at the Zone Hospital and one at each of the two hospitals in Toronto with which it was affiliated. Six units were to be installed in three nursing station–satellite clinic pairs. However, as is often the case in field trials, logistic problems arose at one of the clinic sites, which was more than a mile from the nearest telephone service. The ninth unit was not installed but was held as a replacement to be used when any of the others encountered serious problems.

Carefully designed data collection and database management systems were set up to allow for later comparisons between sites with and without the slow-scan system. Would the number of patient transfers, the number of patient referrals, the relative use of chartered versus scheduled flights for transfers, overnight stays at nursing stations, the duration of stays in hospitals, or the use of X-rays be affected by the technology? All these factors and more would be necessary components in a cost-benefit study of the use of slow-scan television in rural health care.

Although the primary motivation behind the introduction of the system was research, it was emphasized to participants that the equipment would not be withdrawn at the end of the study; if it was useful to them, it would continue to be available.

Quality of implementation is a crucial and too often ignored aspect of any field trial or demonstration project. In this case a key role was that of full-time system coordinator. Responsibilities included training the users of the system, arranging for any necessary repairs, monitoring the data collection process, and maintaining day-to-day liaison with the Zone's health care personnel. The position was held by Helen Acton, a nurse who was highly experienced in working in the remote northern communities and who was well able to empathize with the local health care workers. She played an important part in overcoming the initial hostility that is a common feature of such projects.

The slow-scan television equipment consisted of a black and white television camera with a zoom lens, two black and white monitors, one Colorado Video slow-scan transmitter, one receiver, and a modem. Each of the hospitals had two additional memory units and monitors so that multiple transmissions could be stored and viewed simultaneously (thus, for example, three different X-ray films could be compared). The system provided a picture resolution of 256 by 512 picture elements; it took 78 seconds to transmit each image via the regular telephone lines.

The system was designed to be as simple as possible to use, and all units were identical. Use of the equipment during routine educational and social/therapeutic sessions provided nonstressful opportunities for practice. Minor technical problems were tackled, usually successfully, by local health care personnel with telephone assistance from the system coordinator or, if necessary, from the manufacturer's engineers. Since the equipment was telephonic, the engineers were able to conduct remote diagnostic tests; and in almost all cases, they were able to pinpoint the exact nature of a problem. Major equipment repairs averaged less than one per unit per year during the first five years of operation. The project was successful in avoiding the serious maintenance problems that can easily arise in remote areas which do not have repair facilities.

More than a half of the slow-scan transmissions during the first four years of the system involved X-rays. The image resolution was generally adequate, but this did depend on the experience of the interpreter and on the subject matter of the X-ray film. It took about six months of weekly use for the interpreter's learning curve to level off. Detailed analysis of fine lung structure in chest X-rays posed the most difficulty. Many potential problems could be avoided by use of the zoom capability in the cameras. This, however, depended on the quality of coordination between the sender, who controlled the zoom, and the consultant, who knew where a close-up might be helpful. More of a problem was the quality of the original films. These were taken and developed by the local health care workers, not by trained technicians.

Uses of the system fell into three broad categories: consultations 36 percent, education 37 percent, and social/therapeutic purposes 27 percent. About 55 percent of the transmissions were between the Sioux Lookout Zone Hospital and the northern communities, about 45 percent between the Zone Hospital and one of the two hospitals in Toronto.

Emergency video consultations related mainly to cardiovascular conditions or trauma. For example, a 17-year-old man received a stab wound in the chest. Clearly there was damage to his chest wall and lung, but did he have trauma to his heart or diaphragm as well? The best immediate management of his case depended on the specific type and extent of his injuries. The slow-scan system allowed a combined radiological, cardiovascular, and surgical consultation to take place.

Most consultations were less dramatic. A typical example, occurring at one of the nursing stations, involved a man who had a fracture-dislocation of an index finger. The next visit to the station by a physician was not due for another week. The doctor at the Zone Hospital received an X-ray image of the finger and instructed the nurse on how to reduce the dislocation. After she had done this, she sent another X-ray to check that adequate alignment had been obtained. The doctor then transmitted a page from a textbook showing how to stabilize and cast the finger. On his visit a week later, he satisfied himself that healing was taking place appropriately.

The system has been in regular use for educational purposes as well. For example, medical routine rounds have been conducted from Toronto for physicians at the Zone Hospital and radiological training has been conducted from the Zone Hospital for staff at the nursing stations. Occasionally the system is used for patient/family education, too, as when a physiotherapist instructed a patient's family (and the local supervising nurse) on how certain exercises should be done; as a result, the patient could be discharged earlier from the Zone Hospital.

The final category of use covers occasions when the system has been used to help overcome the anxiety caused by the hospitalization of a family member. One such case arose with a 13-year-old girl who had never left her own community before. She had to be taken to the Zone Hospital and placed in isolation when it was suspected that she had active infectious tuberculosis. She became depressed, hostile, and angry and would talk to no one. A video communication with her parents, however, helped her to express her feelings to the nurses and doctors and to accept the situation better.

A negative finding in the project was that the slow-scan units in the health aide clinics were hardly ever used. Maybe the reasons had to do with the limited familiarity of the health aides with communication technology and the fact that they had no X-ray facilities.

Overall the reactions of the nurses to the system were decidedly positive. Twenty-two of the 30 who responded to a survey after two years of operation stated that they would want to include slow-scan if they were to manage a health care system comparable to that of the Sioux Lookout Zone. However, none of the four physicians based at the Zone Hospital would have wanted to do so. The difference in perception may be due to various causes: Doctors have more extensive training in diagnosis and patient management and are less likely to feel a need for consultation; they are less isolated; and the system placed them in the position of responding to demands made by the nurses, a reversal of traditional roles.

When the capital costs of the equipment were allocated entirely to the remote locations, they came to about $21,000 per site (and would probably be somewhat lower today). Operating costs, mainly long-distance telephone calls, were about $150 per site per month.

Before the experiment it was conceivable that these costs would be offset by financial savings elsewhere, particularly in the costs of stays in the hospital and at the nursing stations, and in the costs of travel. But this was not the case: Costs per patient were slightly higher at those sites with slow-scan television.

To justify the addition of slow-scan television to an audio system from an economic perspective, one would need to assign a sufficient value to educational and social/therapeutic use and to any resulting improvement in health care. The reason that visual communication systems have a hard time demonstrating their cost-effectiveness in rural telemedicine may well be because we can already do so well with appropriate combinations of humbler audio technology and the ultimate in friendly front-ends—the nurse.

Reading

David, W., D.W. Conrath, E.V. Dunn, and C.A. Higgins. *Evaluating Telecommunications Technology in Medicine.* Dedham, Mass.: Artech House, 1983.

Topics for Research or Discussion

━━━━━ What categories might you devise for classifying communication businesses or services? Where, for example, would you include such businesses as cable television and a manufacturer of computers? Where does the telephone business fit into your scheme? After you have developed ideas for a classification scheme, examine a business publication such as the *Wall Street Journal* or *Business Week* and look for examples of communication businesses to include in your analysis; also look in the yellow pages. What generalizations can you make from this analysis about the nature of today's communication businesses and services?

━━━━━ Given the discussion in this chapter of communication activities that exist mainly for social purposes, or the public good, so to speak, how might you contrast their operation with commercially oriented communication businesses? Does being profit-making make a difference in the quality of product? Prepare a paper or a brief discussion on the contrast. Use public versus commercial broadcasting or some other example.

━━━━━━ There is considerable speculation on the future of the entertainment business, including theaters, television in the home, cassettes, cable television, direct broadcast satellite, and the like. What do you think the next decade holds in terms of leisure habits? Will content tastes change? Will entertainment be a growth business? Give reasons for your answers.

━━━━━━ Locate the most recent issue of *Business Week* that contains the quarterly earning reports for the major companies in the United States. Examine the current income, profits, and return on equity of major communications companies, compared with, for example, industrial companies. Generally, how does the communication business compare with other businesses? Examine differences between computer companies and the major communications services companies, for example, telephone or publishing.

━━━━━━ The divestiture of the American Telephone and Telegraph Company is a major milestone in the recent history of communications in the United States. Go to indices of major financial papers (e.g., *The New York Times, Wall Street Journal,* and *Business Week*) for early 1984, where there will be many articles appearing on divestiture. Next, examine the same sources for about a year later. What were some of the unexpected early results of divestiture? What are general contrasts in the financial performance of the remaining AT&T units, compared with the regional telephone companies? Have the same trends continued today?

References and Further Readings

Albertson, L. "Telecommunication as a Travel Substitute: Some Psychological, Organizational and Social Aspects." *Journal of Communication* 27(1977):32–43.

Bell, D. *The Coming of Post-Industrial Society.* New York: Basic Books, 1976.

Campbell, J., and H. Thomas. "The Videotext Marketplace: A Theory of Evolution." *Telecommunications Policy* 5(1981):111–120.

Carey, J. "Videotex: The Past as Prologue." *Journal of Communication* 32(1982):80–87.

Case, D., M. Chen, H. Daley, J. Kim, N. Mishra, W. Paisley, R.E. Rice, and E.M. Rogers. *Stanford Evaluation of the Green Thumb Box Experimental Videotext Project.* Stanford, Calif.: Institute for Communication Research, 1981.

Champness, B., and M. deAlberdi. "Measuring Subjective Reactions to Teletext Page Design." NSF Grant DAR-7924489-A02. New York: Alternate Media Center, New York University, 1981.

Compaine, B.M. *The Newspaper Industry in the 1980s: An Assessment of Economics and Technology.* White Plains, N.Y.: Knowledge Industry Publications, 1980.

Connelly, T. "TAFT Broadcasting Company Findings from Cincinnati Teletext Experiment." In *Videotext '83 Proceedings.* New York: London Online, 1983, 139–146.

Cummings, M. "Medical Information Services: For Public Good or Private Profit?" *The Information Society: An International Journal* 1(1982):249–260.

Danziger, J., W. Dutton, R. Kling, and K. Kraemer. *Computers and Politics: High Technology in American Local Government.* New York: Columbia University Press, 1982.

Dordick, H., H. Bradley, and B. Nanus. *The Emerging Network Marketplace.* Norwood, N.J.: Ablex, 1981.

Dordick, H.S., H.C. Bradley, B. Nanus, and T.H. Martin. "Network Information Services: The Emergence of an Industry." *Telecommunications* (September 1979):217–234.

Drewalowski, H. "German Videotext Experiment Finds Wide Consumer Approval." *Direct Marketing* 36(1983):40–41.

Easton, A. "Viewdata—A Product in Search of a Market?" *Telecommunications Policy* 3(1980):221–225.

Elton, M., and J. Carey. "Computerizing Information: Consumer Reactions to Teletext." *Journal of Communication* 33(1983):162–173.

Feeley, J. "Telidon Trials and Operations—Results, Experiences, and Future Trends." In *Videotex '83 Proceedings.* New York: London Online, 1983, 73–84.

Foster, J., and M. Bruce. "Looking for Entries in Videotex Tables: A Comparison of Four Color Formats." *Journal of Applied Psychology* 67(1982):611–615.

Gaffner, H. "What Is This Thing Called Videotex." In *Videotex '83 Proceedings.* New York: London Online, 1983, 1–16.

Gold, E. "Attitudes to Intercity Travel Substitution." *Telecommunications Policy* 3(1979):88–104.

Goldmark, P.C. "Tomorrow We Will Communicate to Our Jobs." *The Futurist* 6(1972):35–42.

Griffiths, J.M. "The Value of Information and Related Systems, Products, and Services." Vol. 17, *Annual Review of Information Science and Technology,* edited by M. Williams. White Plains, N.Y.: Knowledge Industry Publications, 1982, 269–284.

Hayes, L. "Satellite Technology: Impact on Corporate Publishing." *Graphic Arts Monthly And The Printing Industry* 57(February 1985):S48–58.

InterMedia. "Videotex: Words on the TV Screen," 7:3: entire issue (1979).

Johns, David. "Newspaper Uses of Satellite Technology." Freedom of Information Center, Report No. FOI-847, Columbia, Mo. January 1984.

Kagan, P. "Videotex America Signs Chronicle Publishing." Kagan Newsletter Electronic Publisher, February 28 1983, p.5.

Knight, F.S., H.E. Horn, and N.J. Jesuale. *Telecommunications for Local Government*. Washington, D.C.: ICMA, 1982.

Maloney, E. "Green Thumb Farm Info Project Takes Root in State of Kentucky." *Microcomputing* 25(1982):341.

Marketing News. "Most Consumers Find Videotext Services Useful: Half Would Pay, Survey Reveals." 1982, 16, 11, November 26, 8, 12.

Neustadt, R. *The Birth of Electronic Publishing: Legal and Economic Issues in Telephone, Cable and Over-The-Air Teletext and Videotext*. White Plains, N.Y.: Knowledge Industry Publications, 1982.

Nilles, J., F. Carlson, P. Gray, and G. Hanneman. *The Telecommunication–Transportation Tradeoff*. New York: John Wiley, 1976.

Paisley, W. "Computerizing Information: Lessons of a Videotext Trial." *Journal of Communication* 33(1983):153–161.

Parker, E. "Implications of New Information Technology." *Public Opinion Quarterly* 37(1973):590–600.

Parker, E. and H. Hudson. "Medical Communication in Alaska by Satellite." *New England Journal of Medicine* 289(1973):1351–1356.

Pool, I. de Sola. "The Culture of Electronic Print." *Daedalus* 3(1982):17–32.

Pool, I. de Sola., ed. *Talking Back: Citizen Feedback and Cable Technology*. Cambridge, Mass.: MIT Press, 1973.

Porat, M. "Communication Policy in an Information Society." In *Communications for Tomorrow: Policy Perspectives for the 1980s,* edited by G. Robinson. New York: Praeger, 1978, 3–60.

Ragland, J., and P. Warner. "Green Thumb National Pilot Test." *EDUCOM Bulletin* (Fall 1981):16–19.

Rice, R.E., and W. Paisley. "The Green Thumb Videotext Experiment: Evaluation and Policy Implications." *Telecommunications Policy* 6(1982):223–236.

Smith, D. "Info City." *New York Magazine* (February 9, 1981): 24–29.

Smith, R. *The Wired Nation: Cable TV: The Electronic Communications Highway*. New York: Harper & Row, 1972.

Turoff, M., and S.R. Hiltz. "The Electronic Journal: A Progress Report." *Journal of the American Society for Information Science* 33(1982):195–202.

Tydeman, J., H. Lipinski, R. Adler, M. Nyhan, and L. Zwimpfer. *Teletext and Videotex in the United States*. New York: McGraw-Hill, 1982.

Tyler, M. "Videotext, Prestel and Teletext: The Economics and Politics of Some Electronic Publishing Media." *Telecommunications Policy* 3(1979):37–51.

Webster, F., and K. Robins. "Mass Communications and Information Technology." *The Socialist Register* (1979):285–316.

Wigand, R. "Direct Satellite Broadcasting: Selected Social Implications." Vol. 6, *Communication Yearbook,* edited by M. Burgoon. Beverly Hills, Calif.: Sage, 1982, 250–288.

Wigand, R. "The Direct Satellite Connection: Definitions and Prospects." *Journal of Communication* 30(1980):140–146.

Woolfe, R. *Videotex: The New Television–Telephone Information Services.* Philadelphia: Heyden and Son, 1980.

Media and Computers in Education

Use of technology for instruction has long been a dream of many educational planners and researchers. Yet despite a quarter century of experience with instructional television, it has had little influence on our schools. Computers still hold promise for improving productivity in the classroom—especially with the microcomputer revolution—but the findings remain incomplete and not altogether promising.

TOPICAL OUTLINE

Instructional Television: Promises and Pitfalls

Computers in the Schools

   A Likely Fit
   What Can a Microcomputer Teach (or Learn)?
   Computer Literacy
   Microcomputer Adoption by Schools

Overcoming Barriers to Implementation

INSTRUCTIONAL TELEVISION: PROMISES AND PITFALLS

On the surface, it would seem that education is the ideal environment for the implementation of communication and computing technologies. When the first experiments in educational television were undertaken in the late 1950s, many felt that this new medium could bring instructional materials of the highest quality to every classroom in the nation. This dream has never been fulfilled and probably never will be. The history of instructional television provides generalizations concerning our inabilities for implementing technology in education:

- An attitude long held by many teachers is that large investments in instructional television would threaten their job security.
- Parents who feel that their children watch too many hours of television at home object to having television used in school.
- No unified, efficient network exists for the distribution of television signals to schools or for the distribution of recorded materials. Networks exist, but none effectively serves all schools.
- Curricula have become less flexible and adaptable to the use of new instructional media as school districts, and especially states, have mandated instructional requirements and materials. A local school has less opportunity for innovative use of technology.
- School administrators, including state agencies, have a history of resistance to change. Effective integration of television (and later we will note, computers) requires extensive curriculum change. There is a lack of incentive (and often funds) for making such changes.

As a consequence of these and other factors, educational television is largely a failure as a school-implemented technology in the United States. This is not to say, however, that outstanding programs have not been produced or that there are not superior examples of implementation to be observed on an ad hoc basis.

The major example of success has been the work of the Children's Television Workshop, best known for *Sesame Street,* but equally successful with *The Electric Company* and *3-2-1 CONTACT.* The report by Milton Chen in this chapter describes the meticulous research efforts underlying the development of the *CONTACT* program.

COMPUTERS IN THE SCHOOLS

A Likely Fit

Once available only to a small cadre of engineers and scientists, the computer is now a play and learning tool for children as young as four or five years of age. The latest developments in programs (including videogames) literally allow a child to control images, actions, and alternatives in a visual representation. Computers expand the possibilities for experiencing and even controlling unnatural dimensions of time, space, and speed. (See the report by psychologist Patricia Greenfield on the cognitive advantages of such characteristics.)

The computer holds many implications for the educational, if not social, development of the children of the information age. Indeed, the microcomputer seems to be the most popular topic in the realm of instructional technology, which has a literal as well as figurative fit to life in the information age. (References at the end of the chapter include a number of excellent survey documents.)

For several reasons the computer has been a special, as well as contemporary, topic regarding educational technology in the schools. It has been contemporary in the sense that from the end of the 1970s and continuing well into the 1980s, most of the schools in this country, as well as those in other industrialized countries, have been under pressure to implement computing into classrooms. Part of the reason for this has been the virtual explosion of the microcomputer into the home and office, coupled with a question: Why not the school?

Microcomputers are also a technology that beginning in the early 1980s was increasingly familiar to most children (sometimes more so than to their teachers!). There has also been the call for computer literacy, often involving definitions as a necessary part of the modern school curriculum. In a more general sense, if we were to train our children to cope with computers in the so-called information age, why not use computers as part of the instructional process? The special status of the computer as an instructional technology is that it simultaneously incorporates many features that have long been presumed important to the instructional process—namely, the ability to combine sight, sound, movement, color, and self-pacing, as well as having the qualities of a truly interactive medium. Surely, then, there have been reasons to believe that the computer is a promising instructional technology.

In broader contexts, there has been pressure concerning the quality as well as the status of instructional productivity in our schools. In the same period that blue ribbon committees published such warnings as *A Nation At Risk* (1984), the question is naturally raised whether the microcomputer might offer some technological fix for the problem. If, for example, office productivity can be visibly increased, why not attempt similar applications in the instructional environment?

In this section we review some of the broader generalizations regarding computers in the schools, including observations on various circumstances for implementation and a glimpse of what is known about the instructional software environment, and we conclude with some larger-scale concepts of the computer in the lives of our children.

What Can a Microcomputer Teach (or Learn)?

Most people assume that a computer can teach something. This might be simple math, word lists, shapes, sizes, rules, counting, and so on. But immediate evidence that the computer can be of great advantage in these areas is hard to find. Even if there is specific evidence of an advantageous application, there is the question of whether the enthusiasm for computer use will be maintained, or whether it will be cost-beneficial. Also, there is some bias in the literature (and, we suspect, in teachers' attitudes) that computers are good mainly for teaching math.

Table 11.1 *Instructional Uses of Computers*

Problem solver: The student programs the computer to solve problems (e.g., to do math problems) and thereby learns the process.

Information source: Computer files hold information to aid students in research and problem solving.

Simulator: The computer models a situation or set of actions for the student (for example: How many different shapes can you construct with these five lines? What happens when you add an electrical spark to a mixture of oxygen and hydrogen?)

Grab-bag: The computer gives a student images, numbers, letters, shapes, or anything he would like to experiment with.

Game machines: The computer transforms instructional materials into solo or multiperson games that have learning consequences.

Programming: Both instructional software and practice help students learn programming.

One criticism is that microcomputers are often used as nothing more than electronic flashcards. (Computers can flash items to which the student responds, as in paired-associates learning, spelling, and math drills.) Granted, small computers are used too often in a simple capacity. We need to examine the instructional advantages of the computer medium rather than compare it with simple extensions of other media. One is reminded of early studies of instructional television, in which the comparison was often made between a televised and a live lecture. Little thought was given to capitalizing on the advantages that televised instruction might have over its counterpart.

Before examining other uses, we should note several points about flashcard applications of computers. They are inexpensive to purchase and easy to operate, and with a little instruction most students can program them themselves. We must recognize, too, that drill has traditionally been a part of our instructional strategies, and for certain types of subject matter (e.g., spelling conventions and multiplication tables) there may be no superior teaching exercise.

But what are the alternatives? In a variety of contexts, the computer "as a teacher" can facilitate student learning. The most popular of these applications are summarized in Table 11.1.

Why not "teach the computer"? One of the most promising instructional uses for computers is their application by students as tools for problem-solving: teaching the computer rather than vice versa. In the problem-solver use, a youngster can challenge the computer to do long division by simulating the way humans do it. In doing arithmetic with fractions, a student may require that results be in whole numbers and fractions reduced to lowest terms, rather than in the decimal form a computer will normally yield. In the language arts, a child can be challenged to teach a computer spelling conventions—for instance, the addition of -s, -es, -ez, and irregular plural endings.

As one deals with younger and younger children, techniques for programming may become more complicated than the problem to be solved. Overcoming the difficulties is one advantage of a programming language like LOGO, which is purportedly easier for a child to learn than BASIC. Although LOGO was developed on a mainframe computer system by MIT mathematician Seymour Papert and his colleagues, it has been adapted to several brands of microcomputers. Although these adaptations do not represent all the nuances or advantages of the original LOGO, most have captured useful aspects of it so that it can be made easily available in schools. In fact, we are witnessing a rapid accumulation of reports of demonstration and use of LOGO in practical educational programs. LOGO is an adaptation that brings the computer as a paper-and-pencil-like problem-solving tool into the hands of younger children. Its developers have suggested that it presents an entire "learning environment." Also, we might add, it results more in the child teaching the computer than vice versa. This may be one of the most significant trends today in the uses of microcomputers by youngsters.

Computer Literacy

Computer literacy is a term with as many meanings as proponents. One popular position equates computer literacy with the symbolic tools of language—especially in analogy to the language of mathematics. In this reasoning, a programming language gives a child a system for expressing problems and the algorithms for their solutions. For example, the problem of when two approaching trains will meet can be expressed in an answer-yielding computer program.

There are compelling reasons to question such a specific definition of computer literacy. No programming language is universal as is mathematics; the difference between, say, BASIC and Pascal as computer languages is striking. Moreover, as much as a youngster might benefit intellectually from learning a programming language, including LOGO, it is debatable whether this sufficiently encompasses the assumption that the child can then use computers for problem solving. There is an irony in the trend to make computer language resemble natural language so that people can more easily become computer literate, because computer design is moving in the direction of making machines more "people literate." If the designers ultimately succeed, the user will need to know very little about computers to operate them.

Perhaps definitions of computer literacy that take into account the user's needs for computer applications are preferable. Here *literacy* would take on multiple definitions, some reflecting more on social utility than on direct problem-solving. We must admit that in its strictest definition (ability to read and write), *literacy* is a misleading term as applied to many of the skills involved with computers. Instead, our relative needs for different users of computers are as much qualities of knowledge and attitude as they are skills, and this may be relevant to the broader goal of educating children in the information age. The computer generation will be best educated if they can integrate the technological revolution into fulfilling their differing cultural, social, and psychological needs. Indeed, given the advances in programming languages, artificial intelligence, and the next generation of computers, learning to program in a simple language may already be obsolete.

Microcomputer Adoption by Schools

In the initial phase of the research project on microcomputers in the schools reported by Williams and Williams (1984, 1985), two generalizations are paramount. First, when microcomputers were introduced into the schools that these researchers studied in Southern California, there was something unique about the implementation scenario of almost every school. For example, in one school the math teacher introduced computers by bringing in her own Apple II, whereas in another district, the school board mandated a broad pattern of implementation. The second generalization is that once the computers were settled into the school, the selection and purchase of educational programs (i.e., software or course ware) was almost always the next challenge.

The implementation phase of the research involved visits to 15 schools, 12 of which were eventually summarized as "implementation scenarios" in the Williams & Williams (1984) monograph. Theoretically, there were more cases of horizontal diffusion than in the usual introductions of practices or equipment into public schools. The example of the math teacher fits into this category, as did situations in which, for example, teachers of gifted or remedial programs installed computers, or parents raised money with the expectation of installing several computers in a school, or the president of a PTA (trained in computer science) embarked on a one-person campaign to introduce computers into selected classrooms. In most such cases, the machines simply appeared on the scene, and diffusion took place when other teachers observed their use and decided to acquire them for their own classrooms.

This is in contrast to scenarios in which decisions and plans were made at some higher point in the administrative hierarchy of a school. In some cases, a principal or a particular office in a school district was the innovator, or the acquisition/implementation of computers became a policy set down by the school board itself.

It is not unusual that implementation involves differing scenarios, with varying problems and potential for success. In the cases of horizontal diffusion, although the initial implementation of the computer might have been done with a reasonably clear idea of its application (by the person who brought the machine to school), those who eventually were encouraged to use computers themselves were often less clear on what to expect of them or how to integrate them into the curriculum. In such cases of horizontal diffusion, the problem was less one of enthusiasm or acceptance as it was of efficient acquisition and coordination of resources.

The cases of vertical diffusion, although more clearly orchestrated because planning and implementation were done by the administrative hierarchy, had problems of teacher acceptance, probably more than administrators anticipated. Training teachers to adopt and use computers was a far greater initial expense than buying the machines.

OVERCOMING BARRIERS TO IMPLEMENTATION

Will schools ever become major adopters of communication technologies? Probably not if left to the inventiveness of their personnel. There have been too many barriers to implementation. On the other hand, there are at least two hopeful possibilities.

First, there is an evolving population of students who are coming to school with prior experience with computers, gained outside of school. The recognition by parents that computer literacy may be an important requisite for careers in the information age is now putting pressure on schools to formalize their computer offerings. Although such offerings are more often in terms of "how to" classes in computer familiarization, programming, and applications such as word processing—rather than actually using the computer as a major instructional tool—the rising level of computer expectations may force the expansion of computer integration into the instructional process.

Second, our public schools are facing a growing fiscal crisis. They are victims of an inflationary period in the American economy, but also they have failed to keep pace with economic changes by increasing their productivity. Unlike many businesses, schools carry on their activities much the same as they did 50 years ago. (Compare them with the modern office, the medical profession, or manufacturing.) This pressure to improve productivity may drive schools into reconsidering how communication technologies can be implemented to enhance their effectiveness.

It is imperative that researchers develop improved means for studying the cost-effectiveness of implementing technology in education. We have mostly researched instructional television or computing in terms of student gains rather than the overall effectiveness of the organizational endeavor. When that effectiveness is given higher priority, those who control the purse strings of the schools will be more agreeable to the adoption of technology. It may require fiscal crises to overcome the traditional barriers to implementation.

REPORT

Evaluating 3-2-1 CONTACT

Milton Chen
Harvard Graduate School of Education

The program *3-2-1 CONTACT,* a television series about science and technology for 8- to 12-year-olds, has been broadcast on PBS since 1980. In 1984, the series was awarded five-year support by the National Science Foundation, joining the Department of Education, Corporation for Public Broadcasting, and other funders as part of a national effort to improve science and math education in the United States. *CONTACT* presents science as a fascinating and enjoyable activity open to participation throughout their lives. Its fourth season of 20 shows began broadcasting in the fall of 1985.

Produced by the Children's Television Workshop (CTW), *CONTACT* has four young hosts who present science in minidocumentaries, animation, and studio demonstrations. The series is designed for viewing in both home and school settings. Each week, the five shows are organized by a scientific theme, such as senses, space, or farms. The series also includes a dramatic serial, "The Bloodhound Gang," which involves three young detectives who solve mysteries by logical thinking.

As with CTW's earlier productions for children—*Sesame Street* and *The Electric Company*—formative research has played a major role in the development of *3-2-1 CONTACT*. Researchers trained in communication, education, and psychology have worked collaboratively with television producers, filmmakers, and scientists to develop *CONTACT*'s goals, formats, content, and characters. The formative research staff represents the point of view of the target audience of children in every significant project decision. These decisions include not only the design and production of individual segments and shows, but also the organization of promotion and outreach efforts to increase utilization of the series in schools and homes.

The development phase for the first season, which took from 1977 to 1979, included the most extensive program of formative research ever conducted for an educational television series. More than 50 individual formative research studies were carried out with over 10,000 children, with field work in California, Illinois, Mississippi, New York, New Jersey, Virginia, and Tennessee. Studies included evaluation of existing science shows and films as well as individual segments and prototype test shows for the series.

Research staff also analyzed Nielsen ratings for popular shows among the target audience and surveys of science achievement and attitudes. Researchers helped to disseminate information about *CONTACT* through presentations at conferences of science educators, teachers, and media researchers.

Since broadcast of the first season began in 1980, formative research has continued through each season. Studies have focused not only on the appeal and comprehensibility of show material, but also on the exploration of learning effects from viewing of weeks of *3-2-1 CONTACT*.

The role of formative research, however, does not end with a completed research study. Its real value is in the effective communication of research results, insights, and intuitions to the production staff. This communication occurs not only through formal research presentations and written reports, but also in a continuous dialogue with the project's producers and scientists, from the earliest suggestions of topics to final editing and broadcast. Collaboration between producers and researchers is the heart of "the CTW model," and work on *3-2-1 CONTACT* has continued to emphasize its importance.

The formative research has generated not only substantive findings but also a set of research methodologies for the 8- to 12-year-old audience. Over 20 different techniques have been employed, ranging from simple paper-and-pencil instruments to in-depth small-group interviews and content analysis of children's essays about scientists. The research has taken advantage of innovative technology, such as the QUBE interactive cable system in Columbus, Ohio. There, children evaluated *CONTACT* test shows in their own homes.

One of the most valuable and unusual methodologies involved the design of a microcomputer-based system for measuring viewer response. The Program Evaluation and Analysis Computer (PEAC) is a technique based on the early Stanton-Lazarsfeld Program Analyzer. The PEAC system uses an Apple II computer and 40 hand-held viewer response units to collect moment-by-moment reactions to a program. After viewing, the data are transferred from the microprocessors of the response units to the Apple. Special software for data collection and analysis permits results to be displayed immediately using color graphics. The PEAC system improved the ability of the research staff to collect and report data in a timely and effective manner to project staff.

Results from *3-2-1 CONTACT* research have addressed the intersection of children's developmental abilities, knowledge of science, and television viewing preferences. Some findings include

- *Scientific thinking and knowledge.* Most children in the audience are concrete thinkers. They give literal interpretations of scientific phenomena based on their own experiences with the visible physical world. They do not readily make abstractions.

- *Gender differences.* Boys and girls often differ in their reactions to science topics and cast members. Girls are attracted to stories of human relationships, often in family situations and with strong female characters. Boys prefer action/adventure formats with male lead characters.

While segments featuring animals are of great interest to girls, boys are drawn to a wider variety of scientific and technological subjects, such as outer space and computers.

■ *High appeal of plotted drama.* Children of the target age demonstrate a clear preference for plotted drama over the segmented magazine format as a general program type. They are especially drawn to stories of young people striving to achieve skills and respect in an adult world. "The Bloodhound Gang" purposefully incorporates this format and theme into *3-2-1 CONTACT.* The production of the program involves a constant search for new ways of merging instructional content with a dramatic style of presentation.

■ *Active visuals.* Children in this age group process information from television primarily from the visuals and not from the audio track. Unusual or action-filled pictures of phenomena had a strong appeal, such as the world's largest pizza, a massive oil spill, or an erupting volcano.

The fundamental challenge for *3-2-1 CONTACT* has been to present diverse scientific material with which children had little prior experience, using a medium with which they were intimately familiar. While viewers demanded a fast pace and high production values, they also required that the instruction proceed clearly and carefully. Through successive seasons of *3-2-1 CONTACT,* formative research has proven an important voice in the marriage of television and science.

Further Readings

Chen, M. *A Review of Research on the Educational Potential of 3-2-1 CONTACT.* New York: Children's Television Workshop, 1984.

Children's Television Workshop. *International Research Notes.* Issue No. 3 (3-2-1 CONTACT), Spring 1980.

Mielke, K. *Formative Research on Appeal and Comprehension in 3-2-1 CONTACT.* In *Children's Understanding of Television: Research on Attention and Comprehension,* edited by J. Bryant and D. R. Anderson. New York: Academic Press, 1983, 241–263.

Mielke, K., and M. Chen. *Formative Research for 3-2-1 CONTACT: Methods and Insights.* In *Learning from Television,* edited by M. Howe. London: Academic Press, 1983, 31–55.

▌ **REPORT** ▐

Media Implications for Education and Socialization

Patricia M. Greenfield
The University of California at Los Angeles

More than the medium itself, it is the social context and use of a medium that determine the medium's impact on children's ways of thinking. Print in itself is merely a medium for transferring information; it is not a whole set of higher-level thinking skills. Print is probably a less efficient way to convey information, overall, than is television, with its dynamic visual images, which are more easily understood and remembered than are words.

Television should be used more in schools to communicate information. But it should be used with class discussion directed by the teacher. Children's ability to explain what they have seen on television may well depend on the teacher-pupil dialogue that surrounds the presentation. Like print, television and film are not substitutes for human interaction, but must be combined with and enhanced by it.

In terms of education and socialization, one medium's weakness is another medium's strength. While television has its value, the child also needs other experiences. Parents should restrict the amount of television their children watch at home in order to use other media and experiences to foster reflection and imagination. Encouraging children to read will enhance these types of thinking, while radio (or recordings if children's radio is not available) will stimulate imagination.

The skill of being articulate depends upon knowing how to be verbally explicit. This habit or skill is promoted more by the verbal media of print and radio, less by the audiovisual medium of television. Television seems to promote the use of nonverbal communication, which is also important. Again, a combination of media is desirable.

Because television is so powerful as a learning tool, it is all the more important that children be exposed to high-quality programming that (1) does not go beyond their emotional maturity and (2) provides fantasy or fact that will be useful, not detrimental, to life beyond the television set. How to improve the quality of television programming is a crucial topic, but one that is beyond the scope of this book. But as this and earlier chapters have indicated, parents can do much to improve the effects of television by being selective about what shows children watch and by dis-

Excerpted by permission of the author and publisher from Patricia M. Greenfield, *Mind and Media: The Effects of Television, Video Games, and Computers.* Cambridge, Mass.: Harvard University Press. © 1984 by Patricia Marks Greenfield.

cussing programs to encourage the children to watch critically and thought-fully.

Bringing the electronic media into the schools could capitalize on the strong motivational qualities that these media have for children. Many children who are turned off by school are not turned off by one or another of the electronic media—quite the opposite. An educational system that capitalized on this motivation would have a chance of much greater success. I think it would also make education seem more tied to the "real world," where the importance of the electronic media relative to print is probably the reverse of their relative importance in the world of the school.

Each medium has its own profile of cognitive advantages and disadvantages, and each medium can be used to enhance the impact of the others. In short, to return to Marshall McLuhan, each medium is its own message. The cognitive message of print is the opportunity for reflection. Print and radio share the message of imagination, articulateness, and serial processing. The messages of television and film are an audiovisual style of communication (similar to that of face-to-face communication) and skill in interpreting the two-dimensional representation of movement and space. It may be that television and videogames share the cognitive message of parallel processing. Finally, videogames and computers add the message of interactive learning and the experience of complex interacting variables. The computer is such an open-ended and flexible medium that it also shares messages with many of the media that preceded it. It is interactive like face-to-face communication; it can be a carrier for print, as in word processing; it can be used to program the animated graphics of television or film. It seems too early to say what its final effect on human consciousness will be.

The set of cognitive messages delivered by a particular medium, is, at least in a metaphoric sense, the consciousness created by that medium. It would be a mistake, it seems to me, to become too entrenched in the message of any one medium. Each cognitive message has its own special value.

Educators (myself included) have a tendency to be literary snobs, regretting the passing of an old order in which people knew how to read and write. This attitude has prevented us from seeing the revolutionary promise of the electronic media: They give new cognitive possibilities to disadvantaged groups, and they have the potential to enrich and diversify educational experience for everyone.

Society is also in direct need of the skills that are developed through experience with the electronic media. Already, most people receive most of their information from television, not from print. Feature films provide children's most universal cultural experiences. Thus the need for sophisticated viewing skills is great. Our automobiles are electronic audio environments. Videogames have been lucrative entertainment industries. Computers are inside many items in our everyday environment and are spreading into homes by leaps and bounds. Most of tomorrow's occupations will involve computers in one form or another, and videogames will be most children's first experience in interacting with a computer.

It is not clear how helpful today's television viewing will be for tomorrow's jobs. But society does have a need for people with sophisticated visual skills. E. S. Ferguson pointed out that the language of technology is basically nonverbal and that people involved in technology need to be able to think in terms of images. He said that engineering schools are biased toward educating students to analyze systems using numbers rather than visual images. This bias has produced a lack of people who have the skills to deal with the complexities of real machines and materials.

This bias toward the type of symbol systems used in the medium of print is not limited to engineering schools but is rife in our entire educational system. The time has come to remove this bias and treat the various media as equal, so that our educational system will reflect the messages of those media with which children and adults spend a large part of their lives.

REPORT

Challenges of Educational Software

Victoria Williams
Lake Travis Independent School District; Austin, Texas

Our research project on microcomputers in selected Southern California schools (Williams and Williams, 1984; 1985) involved two phases. The first was to assess the implementation of computers themselves—who obtained them, where they were first used, what were expectations? The second phase involved study of the selection and implementation of educational software. The following generalizations (taken from Williams and Williams, 1985) were drawn from interviews with teachers, administrators, and students from 15 schools in Southern California.

1. Planning and Selection

- Selection of programs has tended to be equipment (hardware) driven because computers are commonly purchased before software. In the long run, it would be much more desirable for software selection to be based upon curriculum needs than constrained by the initial selection (or donation) of computers.

■ There is an existing mismatch between the marketing and sales of educational software and how schools prefer to make acquisitions. Whereas software manufacturers wish to sell large numbers of noncopyable programs to the broadest possible market, educators need a few programs of high quality and duplicates of specialized programs that fit the curriculum needs of individual classroom teachers. The typical distribution of programs via mass market retail outlets is contrary to the educator's need for personal sales attention: previewing, duplication, and follow up assistance.

■ There are often only a few people in a school district who are influential in the selection of software, and their expertise may not be sensitive to the curriculum needs of specific classroom teachers, nor always available to these teachers when they need it.

■ Many teachers do not know how to find the software that they need, nor even what their alternatives are. Mostly, they locate software by word-of-mouth, workshops, in-service, or seeing a successful example. The most widely known software is not necessarily the best. Annotated software bibliographies are not widely available to classroom teachers. Few teachers seem interested in creating their own software or have the training or time for doing so.

■ Funding remains a major problem in the acquisition of software, especially when budgets have concentrated on equipment. It is not unusual to see software purchased with funds diverted from textbook, aide, or other budgets. Parent, PTA, and business groups sometimes assist with software funding. With duplication of disks so easy but usually illegal, schools are frustrated to have to pay full price for each duplicate of an adopted program.

2. Curriculum Applications

■ Fitting educational software into the curriculum is a major challenge. The most innovative uses of educational software may require substantial curriculum revisions. Software that is compatible with an existing curriculum does not tend to make the best advantage of educational uses of microcomputers. (This software is often of the "drill and fill" type, and ironically, many teachers find it the easiest to use.)

■ Objective information about the effectiveness of educational software is necessary. There is a need to be more specific about how software stimulates or reinforces learning. Specialists in educational computing are calling for more software that promotes discovery and creative thinking and teaches the use of the computer as a problem-solving tool in relation to the curriculum.

■ Teachers are frequently not made aware of what software is available for given curriculum plans. As one district officer phrased it, "When new curriculum objectives are developed, they need to include specific references to instructional aids, including computer software."

■ Various subjects and grade levels need different software. Math and science applications still predominate, with the humanities and arts far less in evidence. In the early elementary grades, software is most commonly used to reinforce basic learning skills. In second through fourth grades, software seems especially effective for introducing new concepts and creating experiences with them. In the upper grades, software can be used effectively as a tool for problem solving, including the students' learning to program. In high school, effective software includes the familiar productivity tools—e.g., word processing, spreadsheets, database management, and graphics. Packages for introducing advanced programming languages are also important. There may be special opportunity, also, for uses of software in vocational training areas and certainly in computer training.

■ More attention needs to be given to such strategic questions as: How can software be designed to take advantage of varying learner styles? What is the potential for software that is used by several students or even groups? How is student and teacher time effectively reallocated when computers are used? Does the software free the teacher for other tasks?

3. Teacher Training

■ A teacher's degree of competence, more than the quality of software, is often the reason for the success or failure of microcomputer implementation. The best educational software can be ineffective without proper teacher support. Teacher training is important, not only for microcomputer implementation but relative to software use. In fact, as teachers become familiar with computers, software becomes the most important part of training.

■ Teachers often say that it is difficult for them to gain practical and meaningful experience with new software prior to its selection and use. They need more time for try-out—even in the classroom—than is afforded by workshops or a short-term loan of a program.

■ Better software documentation, including suggestions for curriculum implementation, tips for student motivation, and innovative applications, might lessen the problem of software training.

■ Introducing teachers to software that aids them in their own work (e.g., gradebooks and word processing) encourages instructional implementation.

- Effective training seems best accomplished at school, rather than in outside workshops, and in comfortable group environments. Many teachers do not see the benefits of training that focuses too much on the insides of computers, the wonders of technology, the history of computers, or programming, at the expense of knowing about software and how best to implement it.

4. Attitudes

- Teachers, students, and administrators have somewhat contrasting attitudes about what they seek in software. Administrators want programs that meet broad curriculum goals and will not become obsolete. Teachers seek an effective fit for their classrooms and their personal styles of instruction, as well as programs that focus on well-defined instructional objectives. Students like variety and the ability to control what they will be doing and are most interested in the latest programs.

- Some administrators still have sufficient doubts about the effectiveness of computers in school so that even if machines have been obtained (or donated), they are skeptical about investing significantly in new software. Most administrators prefer software for which there are known standards of effectiveness (seldom available). Administrators' concerns about the security of equipment extend to software, which by its nature is easily lost, ruined, or stolen.

- Computer specialists often express frustration with the incompatibility of programs among various computers. This constrains large investments in software because of the fear that it will become obsolete with new models of computers.

- Frequently expressed teacher attitudes include a preference for software that (1) is self-administering, (2) includes on-screen instructions, (3) is fail-safe (i.e., hard to stop inadvertently), and (4) has clear benefits for classroom learning. Although most software experts and resource staffs advocate avoidance of drill programs, teachers often like them for their simplicity. While students are drawn to game-type programs and "arcade" effects, teachers often question their educational value. Teachers especially dislike software that requires too much of their personal attention for the computer's operation. It is difficult to "fiddle" with an ailing program in a room with 30 children competing for attention.

- Students' frequently expressed attitudes include a desire for software (1) with "new, different, and exciting" qualities, (2) that lets them control what they do, (3) that has arcade game qualities, (4) that includes the best graphics, color, and sound effects, (5) that allows them to move at a comfortable speed, and (6) that lets them try a problem again rather than being wrong. Simple drill programs are soon boring, as is the repetitious "good answer, Sally" feedback from the computer. Some of the more skilled students prefer to use the machines for their own purposes (e.g., copying each other's game programs or working up new programs) rather than for instruction.

■ It is only a first step for parents to be enthusiastic about their children's use of computers in school. The next step is to want to know more about the programs they use and whether the software contributes to education. (They do not want their youngsters spending time with game programs at school.)

■ Some administrators and computer specialists feel that manufacturers of educational software are indifferent about the schools. Their inquiries to them may go unanswered or left to a public relations department.

5. Evaluation

■ Administrators' evaluation of software seems to be divided into three levels: (1) the apparent quality of programs and their documentation, (2) whether the selection process puts the proper software into the classroom, and (3) whether classroom practices use student and teacher time effectively, and the degree to which they are relevant to the curricula.

■ Evaluation is usually made on the basis of the apparent quality of programs. Rather than being formally evaluated, software is chosen mostly informally by word of mouth, although some schools and district offices have questionnaires that are completed by users and then made available for inspection.

■ Program reviews in computer magazines are met with mixed opinions by those in charge of school software, sometimes because there is a feeling that the reviewers have more experience with computers than classrooms or subject matter.

■ Some computer resource specialists suspect that the immediate effectiveness that they witness when they are implementing software often fades when outside attention is no longer involved.

■ Computer specialists say that they would like to see software manufacturers devote more resources to formative evaluation of their product and to make the results available with program documentation.

For further readings, see the references at the end of this chapter.

Topics for Research or Discussion

━━━━━━━ Among the many theories of how children interact with computers is Sherry Turkle's thesis that children anthropomorphize computers or robot toys (that is, they tend to assume that the computers are humanlike). Review Sherry

Turkle's book, *The Second Self* (New York: Simon and Schuster, 1984). Prepare a paper that describes Turkle's thesis and your own opinions on the topic.

■■■■■■ One of the most widely read and influential books on children and microcomputers is Seymour Pappert's *Mindstorms* (New York: Basic Books, 1982). Among other things, Pappert theorizes that the computer assists the child in being able to "make the abstract concrete." Review Pappert's book and then prepare a paper that describes his main views on children and computers. Conclude with your own critical evaluation of these views.

■■■■■■ The more aggressive and socially aware cable television companies often seek strategies for cooperation with civic institutions such as local governments, schools, and hospitals. Identify one such application in your community, then interview local cable officials and officials of that institution as to types of cooperative projects. Prepare a consulting report on your results. (TIP: You may be able to search the Educational Resources Information Center database, which contains entries on cooperative projects involving cable television and education.

■■■■■■ An interesting generalization is that young children who have had little direct experience with computers tend to have attitudes toward them based upon experiences with television or movie portrayals. Interview a dozen or so children in the age range between 4 and 8 years. Inquire into their thoughts about computers, including descriptions of them. Draw implications as to what you believe are the bases for these images and attitudes.

■■■■■■ See if your university library can gain you access to a computerized bibliographic service. (Or perhaps your department or school has direct access.) Conduct a search on a topic of your choice. Bring the print-out to class. Be prepared to discuss the relative advantages and disadvantages of on-line reference services.

References and Further Readings

Anderson, R., D. Klassen, K. Krohn, and P. Smith-Cunnien. *Assessing Computer Literacy. Computer Awareness and Literacy: An Empirical Assessment*. Minneapolis, Minn.: Minnesota Educational Computing Consortium, 1982.

Ault, R. "Putting the Kids in Control." *Classroom Computer News* (November–December 1982): 46–47.

Bardige, A. "The Problem Solving Revolution." *Classroom Computer News* (March 1983): 44–46.

Becker, H. "How Schools Use Microcomputers: First Report from a National Survey." Baltimore, Md.: Center for Social Organization of Schools, Johns

Hopkins University, 1983, paper presented to the American Educational Research Association, Montreal, April 1983.

Becker, H. "Microcomputers in the Classroom: Dreams and Realities." Baltimore, Md.: Center for Social Organization of Schools, Johns Hopkins University, 1982.

Bitter, G.S. "The Road to Computer Literacy." *Electronic Learning* (September 1982): 60–63; (October 1982): 34–68; (November–December 1982): 44–91; (January 1983): 40–48; (February 1983): 54–60.

Bonner, P. "Toward a More Thoughtful Use of Computers in Education." *Personal Computing* (February 1984): 153–61.

Bork, A. "Production Systems for Computer Based Learning." In *Instructional Software: Principles and Perspectives for Design and Use,* edited by D.F. Walker and R.D. Hess. Belmont, Calif.: Wadsworth, 1984.

Bowman R.F. "A Pac-man Theory of Motivation: Tactical Implications for Classroom Instruction." *Educational Technology* (September 1982): 14–16.

Brahan, J., and D. Godfrey. "A Marriage of Convenience: Videotex and Computer Assisted Learning." *Computers and Education,* 6(1982):33–38.

Bretz, R. *Media for Interactive Communication.* Beverly Hills, Calif.: Sage, 1983.

Chen, M., and W. Paisley, eds. *Children and Microcomputers.* Beverly Hills, Calif.: Sage, 1985.

Clarke, P., T. Kline, H. Schumacher, and S. Evans. "Rockford, Ill.: In-Service Training for Teachers." *Journal of Communication* 28(1978):195–201.

Dede, C. "Educational, Social and Ethical Implications of Technological Innovation." *Programmed Learning and Educational Technology* (November 1981): 204–215.

Dirr, P., and R. Pedone. "Uses of Television for Instruction, 1976–1977: Final Report of the School TV Utilization Study." Washington, D.C.: Corporation for Public Broadcasting, 1979.

Emmett, A. "Discovering a New Way to Learn." *Personal Computing* (January 1984): 56–68.

Federico, P.A. "Individual Differences in Cognitive Characteristics and Computer Managed Mastery Learning." *Journal of Computer-based Instruction* 9(1982):10–18.

Gable, A., and C.V. Page. "The Use of Artificial Intelligence Techniques in Computer Assisted Instruction: An Overview." *International Journal of Man-Machine Studies* 12(1980): 259–282.

Gagne, R.M., W. Wager, and A. Rojas. "Planning and Authoring Computer Assisted Instruction Lessons." *Educational Technology* 21(1981):17–26.

Gibbon, S., E. Palmer, and B. Folwes. "Sesame Street, The Electric Company, and Reading." In *Toward a Literate Society,* edited by J. Carroll and J. Chall. New York: McGraw-Hill, 1975.

Greenberg, B. "Television and Role Socialization: An Overview." In *Television and Behavior: Ten Years of Scientific Progress and Implications for the 80s,* edited by D. Pearl, L. Bouthilet, and J. Lazar. Washington, D.C.: National Institute of Mental Health, 1983, 179–190.

Greenfield, P.M. *Mind and Media: The Effects of Television, Video Games, and Computers.* Cambridge, Mass.: Harvard University Press, 1984.

Hechinger, F.M. *About Education: Computer Software Found Weak. New York Magazine* (April 20, 1982):C4.

Hess, R., and I. Miura. "Gender and Socioeconomic Differences in Enrollment in Computer Camps and Classes." Stanford, Calif.: Stanford University School of Education, 1983.

Himmelweit, H., A. Oppenheim, and P. Vince. *Television and the Child.* London: Oxford University Press, 1958.

Johnson, J., and J. Ettema. *Lessons from Freestyle: Creating Presocial Television for Children.* Beverly Hills, Calif.: Sage, 1982.

Kearsley, G. "Videodiscs in Education and Training: The Idea Becomes Reality." *Videodisc/Videotex* 1(1981):208–220.

Lathrop, A., and B. Goodson. *Courseware in the Classroom.* Reading, Mass.: Addison-Wesley, 1983.

Lepper, M. "Microcomputers in Education: Motivational and Social Issues." Stanford, Calif.: Stanford University School of Psychology, 1982.

Lesse, S. "The Preventive Psychiatry of the Future." *The Futurist* 10(1976):228–237.

Leuhrmann, A., and H. Peckham. *Computer Literacy: A Hands On Approach.* New York: McGraw-Hill, 1983.

Levin, J., M. Boruta, and M. Vasconcellos. "Microcomputer-Based Environments for Writing: A Writer's Assistant." In *Classroom Computers and Cognitive Science,* edited by A. Wilkinson. New York: Academic Press, 1983.

Liebert, R., J. Sprafkin, and E. Davidson. *The Early Window: Effects of Television on Children and Youth.* 2d ed. New York: Pergamon, 1982.

Lyle, J., and H. Hoffman. "Children's Use of Television and Other Media." Vol. 5, *Television and Social Behavior,* edited by G. Comstock, A. Rubenstein, and J. Murray. Washington D.C.: U.S. Government Printing Office, 1972, 129–153.

Malone, T. "What Makes Things Fun to Learn? A Study of Intrinsically Motivating Computer Games." Palo Alto, Calif.: Xerox Palo Alto Research Center, 1980.

Malone, T.W. "Toward a Theory of Intrinsically Motivating Instruction." *Cognitive Science* 4(1981):333–369.

Mason, G. "Computerized Reading Instruction: A Review." *Educational Technology* 20(1980):18–22.

Needham, N. "The Impact of Videogames on American Youth." *Today's Education* 71(1982):52–55.

Office of Technology Assessment. *Implications of Electronic Mail and Message Systems for the U.S. Postal Service.* Washington, D.C.: U.S. Government Printing Office, 1982.

Paisley, W., and M. Chen. "Children and Electronic Text: Challenges and Opportunities of the 'New Literacy.'" Stanford, Calif.: Stanford University Institute for Communication Research, 1982.

Paisley, W.J., and M. Chen. "Children and Interactive Media: Exploring the Effects of the Second Electronic Revolution." Vol. 8, *Communication Yearbook,* edited by R. Bostrom. Beverly Hills, Calif.: Sage, 1984.

Palmer, E. "Formative Research in the Production of Television for Children." In *Media and Symbols: The Forms of Expression, Communication and Education,* edited by D. Olson. Chicago: University of Chicago Press, 1974.

Papert, S. *Mindstorms: Children, Computers and Powerful Ideas.* New York: Basic Books, 1980.

Pogrow, S. *Education in the Computer Age.* Beverly Hills, Calif.: Sage, 1983.

Raleigh, C.P. "Give Your Child a Head Start." *Personal Software* (November 1983): 36–43.

Riordan, T. "How to Select Software You Can Trust." *Classroom Computer News* (March 1983): 56–61.

Salomon, G. *Interaction of Media, Cognition and Learning.* San Francisco: Jossey-Bass, 1979.

Salomon, G. "Television Watching and Mental Effort: A Social Psychological View." In *Children's Understanding of Television: Research on Attention and Comprehension,* edited by J. Bryant and D. Anderson. New York: Academic Press, 1983.

Schramm, W., J. Lyle, and E. Parker. *Television in the Lives of Our Children.* Stanford, Calif.: Stanford University Press, 1961.

Sherman, S., and K. Hall. "Preparing the Classroom for Computer-Based Education." *Childhood Education* 59(1983):222–226.

Steffin, S. "The Educator and the Software Publisher: A Critical Relationship." *Technological Horizons in Education Journal* (March 1982): 63–64.

Sugarman, R. "What's New, Teacher? Ask the Computer." *IEEE Spectrum* (September 1978):44–49.

Suppes, P., and M. Morningstar. *Computer Assisted Instruction at Stanford, 1966–68: Data, Models, and Evaluation of the Arithmetic Programs.* New York: Academic Press, 1972.

Torrance, E.P. "Implications of Whole-Brained Theories of Learning and Thinking for Computer-Based Instruction." *Journal of Computer-Based Instruction* 7(1981):99–105.

Turkle, S. *The Second Self: Computers and the Human Spirit*. New York: Simon & Schuster, 1984.

Unwin, D. "The Future Direction of Educational Technology." *Programmed Learning and Educational Technology* (November 1981): 271–273.

Vockell, E.L., and R. H. Rivers. *Instructional Computing for Today's Teachers*. New York: Macmillan, 1984.

Walker, D.F., and R.D. Hess. "Evaluation in Courseware Development." In *Instructional Software: Principles and Perspectives for Design and Use*, edited by D. F. Walker and R. D. Hess. Belmont, Calif.: Wadsworth, 1984.

Wartella, E. "Children and Television: The Development of a Child's Understanding of the Medium." Vol. 1, *Mass Communication Review Yearbook*, edited by G. Wilhoit and H. DeBock. Beverly Hills, Calif.: Sage, 1980.

Wasserman, A.I., and S. Gutz. "The Future of Programming." *Communications of the ACM* 25(1982):196–206.

Wilken, W., and C. Blaschke. "States Need to Rethink Software Strategies." *Classroom Computer News* (May–June 1983):96.

Williams, F., R. LaRose, and F. Frost. *Children, Television and Sex-Role Stereotyping*. New York: Praeger, 1981.

Williams, F., and G. Van Wart. *Carrascolendas: Bilingual Education Through Television*. New York: Praeger, 1974.

Williams, F. and V. Williams. *Growing Up With Computers*. New York: William Morrow, 1983.

Williams, F. and V. Williams. *Microcomputers in Elementary Education*. Belmont, Calif.: Wadsworth, 1984.

Williams, F. and V. Williams. *Success with Educational Software*. New York: Praeger, 1985.

Communication Technology and Economic Development

We have long known that technology and socioeconomic development go hand in hand. This seems particularly true for communication technologies. Whereas scholars such as Wilbur Schramm and Daniel Learner placed the stress on mass media and development, current theorists such as Everett Rogers and Heather Hudson stress the importance of telecommunications and computing technologies, a concept that Herbert Dordick calls the "communication infrastructure." This chapter examines the characteristics of that concept.

TOPICAL OUTLINE

National Investments in Communication Systems

The Concept of Postindustrialism
Stages of Socioeconomic Development
Sociologizing and Economizing Modes

Developmental Theses
Diffusion of Innovations
Technology as Economic Investment

Whither the Information Society?

NATIONAL INVESTMENTS IN COMMUNICATION SYSTEMS

A key thesis of this chapter is that there is a correlation between a nation's economic development and the extent of its communications infrastructure. Table 12.1 provides summary statistics on media and communications technologies existent in major parts of the world. One main generalization from this table, which is common knowledge to students of international communication, is that radio is by far the most widely installed and available medium in the world. No doubt this is because radio is inexpensive to install and operate and does not require literacy; since the advent of the transistor, the price of receivers has fallen within reach of most of the world's population.

In direct contrast to the radio is the telephone, a much more expensive national investment, which is found in large numbers only among the most highly developed, wealthy nations. The correlation between national wealth and the presence of telephones is high—approximately .68 (Hardy, 1980). The examination of the distribution of electronic computers is also noteworthy. Some nations have virtually none and others a monopoly.

Given these contrasts, an important topic for study is *how* the existence of communication technologies is related to national wealth. Put into practical terms: Do national investments in communication technologies contribute to development? In this chapter, we explore several theories about this relationship.

THE CONCEPT OF POSTINDUSTRIALISM

Stages of Socioeconomic Development

In several earlier chapters, we referred to Daniel Bell's concept of postindustrialism, the idea that societies grow through certain stages of economic development. The contrast is made among agrarian or preindustrial, industrial, and postindustrial stages of growth. A preindustrial society gains its wealth from the land or sea; it extracts value from nature by farming, lumbering, mining, or fishing. An industrial society adds value to raw materials by manufacturing them into products. This source of value evolved from the craftsmanship of the Middle Ages to the machines of the industrial revolution. That is to say, a society could generate wealth by employing manufacturing capabilities. A society like Japan, which lacks substantial raw materials yet in turn bases its economy on creating and exporting products made from the materials of others, is a prime example of industrialism, as was England in the last century.

Postindustrialism, the newer concept, is an economy in which information is itself increasingly a source of wealth. A prime example of valued information is the process of research and development (R&D). The results of research can be sold, licensed, traded, or used for profit. Modern examples of R&D are

Table 12.1 *Statistics on World Media*

The following statistics for the year 1981 are drawn from portions of the *1982 Statistical Yearbook (Annuaire statistique)*, published by the United Nations, New York, 1985.

Country	Radio per 1000 inhabitants	Television per 1000 inhabitants	Telephone per 100 inhabitants
AFRICA			
Algeria	204	61	3.3
Egypt	150	40	1.2
Liberia	172	11	−.1
S. Africa	275	70	11.8
NORTH AMERICA			
Canada	1149	489	69.3
Cuba	309	154	4.2
United States	2110	631	78.9
SOUTH AMERICA			
Argentina	748	197	9.8
Brazil	308	122	7.2
Peru	159	49	−.1
ASIA			
Afghanistan	76	3.1	−.1
China	60	5	−.4
India	59	1.8	.5
Israel	253	240	31.3
Japan	688	551	47.9
Saudi Arabia	295	248	11.2
EUROPE			
France	927	361	49.8
Greece	350	160	30.2
Italy	243	390	36.4
Norway	329	298	48.5
United Kingdom	963	411	49.7
OCEANIA			
Australia	1112	380	no data
Samoa	318	19	3.7
USSR	504	306	9.3

research in the high tech industries—microelectronics, biochemistry, and artificial intelligence, to name a few. Other examples of value include information and information-processing devices that enhance existing methods for generating wealth, such as computer-aided design or manufacturing, management information systems, financial networks, and office automation devices. These last examples are less "pure" information (than in R&D) as they are capabilities for gathering, interpreting, storing, or communicating information. They are visible as computers, software, telecommunications networks, and a variety of input/output devices (e.g., terminals and printers); in all, they are the technologies described throughout the present volume.

The hallmark of the postindustrial age is that information technologies are the basic tools for enhancing our human abilities. Just as machines amplified human physical capabilities in the industrial age, information technologies expand human cognitive and communication capabilities in the postindustrial one.

The stages of socioeconomic development are not mutually exclusive. A nation needs raw materials and the products of manufacturing although it be in the midst of postindustrial growth. This overlap reflects two further key generalizations. First, as national economies become specialized, nations become more interdependent for their resources, manufacturing, and information needs. An example of this is the increasing tendency for the manufacturing of electronics to be done offshore from the United States, although the basic R&D is a product of our postindustrial economy. For years, the USSR, while emphasizing the growth of heavy industries, has not been able to avoid the necessity of importing food. The second key generalization is that a new stage of economic development is often based upon services or capabilities delivered to a prior one. For example, the United States is a world leader in agriculture because it applied industrial techniques to farming methods—e.g., the invention of the harvesting machine. In the latter part of this century, information has added greatly to agricultural methods, as in the development of pesticides, fertilizers, and crop rotation methods. In short, although agriculture is characteristically a preindustrial means for creating wealth, modern agriculture is dependent upon manufacturing and information technologies.

It is within this broad concept of socioeconomic stages that we can best examine the role of communication technologies in development.

Sociologizing and Economizing Modes

Another key concept (earlier introduced in Chapter 10) in Bell's postindustrial theory is the distinction between *sociologizing* and *economizing* modes in development. The former, as the root term implies, relates to development that has goals in social benefits. The latter relates mainly to efficiency and, in business, profit. Bell forecasts that sociologizing goals will become increasingly important to Americans in the evolving postindustrial era.

National investments in, or regulation of, communication technologies in the United States well reflect the sociologizing-economizing distinction. The

concept of "affordable and universal" service that underlay telephone regulation in this country is a sociologizing mode, although the Bell system was also
good business. Broadcasting for "public convenience and necessity" is a
socially oriented premise underlying broadcast regulation in this country.
However, that both the telephone and broadcasting industries were organized
as "for profit" business is characteristic of economizing goals.

The current and continuing trend toward deregulation in this country also
has sociologizing-economizing connotations. For both telephony and
broadcasting, the rationale for deregulation is that the businesses will experience accelerated growth and long-range profits. There will be more incentive
for investment. These, of course, are economizing arguments. The sociologizing aspect is that the public will eventually benefit from having more products,
services, and perhaps lower prices.

But what of the world scene? First, it is important to bear in mind that most
of the world's telephony and broadcasting are organized as state agencies or at
least governmentally controlled agencies. They are, a priori, organized along
sociologizing lines, and for most of their existence, little thought had been
given to their economic implications. They were part of the overhead of
governance. And the more authoritarian the political climate, the more important the effective control of these agencies. They have, for the most part, been
political and social tools.

Yet in the years following World War II, as more national attention has
been directed toward modernization, communications media and technologies
have grown in importance. In countries having a primarily sociologizing
approach to communication systems, there is increased interest in the
economizing mode. Thus, it is not unusual to see a trend toward deregulation
or divestiture of state-operated telephone and broadcast systems to attract
private investment and promote growth. France is a current example of this
trend.

DEVELOPMENTAL THESES

Diffusion of Innovations

There are theories that can assist us in conceptualizing the relation between
communication technology and national development. For this discussion, let
us consider two of them. The first, the diffusion of innovations, dating from
the 1950s and 1960s, reflects the use of media for public orientation, education, and politicalization. Although political leaders since the time of ancient
civilizations have respected the power of persuasion, there was growing
interest forty or so years ago in how communication enhanced the flow of new
ideas through a society. As stated as a theory of diffusion of innovations:

> The main elements in the diffusion of new ideas are: (1) an *innovation*,
> (2) which is *communicated* through certain *channels*, (3) *over time*, (4)
> among the members of a *social system*. (Rogers, 1983, p. 35)

The basic idea is not complex (although the process may be). Modernization requires the diffusion of ideas, and a nation's systems of communication are critical to this process. This is a "flow of know-how" theory. Mass media theorists Wilbur Schramm and Daniel Lerner emphasized how broadcasting and newspapers contribute to the process. Sociologist and communication researcher Everett Rogers saw it as a larger process involving the flow of ideas from media to opinion leaders, then spread by interpersonal communication to the population. In either case, to encourage national development, a country should promote the diffusion processes, and media are key factors.

Technology as Economic Investment

In recent years, another view has been gaining attention. This is not so much contrary as it is complementary to the diffusion theory. The concept is that national investment in media, especially telecommunications technologies, promotes economizing effects upon a nation's businesses. That is, the availability of improved communication channels stimulates an economy. One example is the small manufacturer who by advertising broadens the market for his products. Another is the farmer who can benefit from weather information. Still another is the fisherman who can use the telephone to sell a catch before it is delivered. An illustrative study in this area was Hardy's (1980) assessment that a nation's investment in an improved telephone system would bring more economic return than an investment in radio broadcasting. Hudson explores the same concept in her monograph, *The Telephone Comes to the Village* (1984).

This reasoning has also gained prominence in the work of specialists in urban development and land planning (Castells, 1985). Mostly this research has been focused upon the evolution of a high tech style of economic growth. Important questions include: How does the expansion of technical occupations (e.g., scientists and engineers) and investment in technologies (computers and telecommunications) contribute to the growth of an economy? What are the contrasts between regions (e.g., the northeast versus the south in the U.S.) in this component of growth, and why do those contrasts appear? And ultimately, how do these changes affect culture and life styles?

WHITHER THE INFORMATION SOCIETY?

As we approach the end of the 1980s, most countries of the world are asking about their roles in the evolving information economy. Developing countries are looking toward communication and computing investments as a strategy for hastening economic development. New industrialized countries like Taiwan and Korea are competing with Japan for world markets, while less developed countries like Indonesia and the Philippine Islands have their eyes on competing with Taiwan and Korea.

Even the mature industrialized countries of Northern Europe are seeking growth strategies based upon finding a niche in the new economy. As Spain entered the Common Market in 1986, it had already embarked on an aggressive policy of developing liaisons with multinational high technology companies wishing to locate manufacturing facilities on the continent. And France has committed itself to a policy of information technology growth based on the establishment of national institutes and projects for the diffusion of new technologies and deregulation activities designed to attract investment.

Finally, the revolution in communication technologies is realigning the competitive positions of the United States and the Soviet Union. The Soviets realize that they must improve the productivity of their traditional industries and that this requires new technologies and management. Their landmark growth as an industrial power following World War II has lately declined as many of their factories have become obsolete in the information age. The move toward high technology is now high on their agenda for reindustrialization.

The United States, in contrast, has recognized and attempted to capitalize upon the new technologies as a priority for economic growth. But a major portion of this investment reflects military-related expenditure, a situation that directly bears upon the "sociologizing versus economizing" dichotomy posed by Daniel Bell (1976). Whether America can direct her efforts in communication technology and national development toward a positive human goal rather than toward destruction is the challenge of our age. This challenge defines the dramatic options offered as answers to the question that concludes this chapter: *Whither the information society?*

REPORT

Communication Technology in the Third World

Everett M. Rogers
The University of Southern California

Three-quarters of the world's population lives in hunger and poverty, suffers from high rates of infant mortality, and is unable to read and write. These people are concentrated in the Third World countries of Latin America, Africa, and Asia and live in rural villages and urban slums. They and their governments are trying to improve their lives through programs of economic, social, and educational development.

How can new communication technologies such as computers and satellites facilitate these efforts? The answer begins with the transistor radio revolution of the 1960s. Due to advances in microelectronics, radios suddenly became miniaturized, much cheaper, and could use batteries as a power source. I was living in Colombia in 1963 and 64 when the transistor radio came to that country. A radio set dropped in price to about $4 (U.S.), which even peasants could afford. Shortly, one could hear music and news in villages and in slums. The mass population of Colombia was thus brought into the mass media audience. The special advantages of the transistor radio for the Third World were its affordable price and its ability to overcome the illiteracy barrier.

The 1960s were a very upbeat period in thinking about the potential of the mass media in bringing about development in the Third World. The media were expected to act as "magic multipliers" in providing useful information to the public. Our expectations were soon dashed by the experience with instructional television in such developing nations as El Salvador, Samoa, Colombia, and the Ivory Coast. The basic idea was to use television to multiply access to the teaching abilities of outstanding instructors, so that every classroom in the nation could benefit from superior pedagogy. Television was to be a window to the world for the students in Third World nations. So instructional television broadcasting facilities were constructed, at considerable cost. Later, evaluation studies of instructional television showed that only small gains were achieved in student learning as a result of this technology.

In 1945, Arthur C. Clarke, then a young pilot in the RAF, wrote an article, published in a radio journal, calling for television broadcasting from geosynchronous satellites. He realized that if a satellite was positioned 22,300 miles from the earth, its orbit would remain in a fixed position relative to the earth's surface. Thus, a broadcasting satellite could function as the tallest TV tower imaginable. From a single satellite, a television signal could reach approximately one-third of the earth's surface. Clarke realized the special advantages of television satellites for Third World nations, especially those with a large land area, mountains, or other obstacles to conventional television broadcasting.

One of the first large-scale trials of Clarke's idea was the Satellite Instructional Television Experiment (SITE) in India in the mid-1970s. This project involved 2400 villages, each equipped with a TV set for community viewing and a small receiving antenna. Programs were broadcast that dealt with agriculture, family planning, health, and nationalism. The results of SITE showed that the hardware technology of satellite television worked satisfactorily, but the broadcasts had only a limited effect on the village audiences for which they were intended.

During the 1980s, India has pushed on with other attempts to extend television to the entire countryside. A television message is now beamed via satellite to several hundred local TV stations, each of which rebroadcasts the message over a radius of 20 to 25 miles. About 80 to 90 percent of India's population are thus able to receive TV signals. The limitation to wider TV exposure is the cost and availability of TV sets.

In the People's Republic of China today, a rapid rate of adoption of television is underway. A 1983 sample survey of households in the Beijing metropolitan area showed that 92 percent of adults regularly viewed television. The expansion of television in China began only after 1976, but by 1983 over 20 percent of all Chinese households had purchased TV sets (71 percent in urban areas).

So in many Third World countries, thanks in part to satellite technology, a large share of the public now is getting television exposure.

Microcomputers are another promising technology, but their impact has been very limited in the Third World. A number of experiments are underway, designed to test the usefulness of microcomputers for education in schools, for administrative record-keeping in government agencies, and for diagnostic purposes in health clinics. Computers are still costly for poor countries, and much work remains to be done to determine how useful they can be in the Third World.

One of the important lessons learned about the media's role in development is that the content must be appropriate to the audience or development will not occur. For example, much of the radio and television programming is music and other forms of entertainment, without much content devoted to improved agriculture, better health, or education. In India the most popular TV programming is old movie films. Media advertising shows a good life to poor people and so may lead them to desire more income and better food and housing. But the media do not provide information that can fulfill these desires. The result may be frustration.

Several attempts are being made to provide media content that is more relevant to development. Starting in 1978, Televisa (Mexico's commercial TV system) began broadcasting a series of soap operas with a development theme. The first year emphasized adult literacy, the second year family planning, and a later series stressed good family relationships. These soap operas earned the highest audience ratings on Mexican television, showing that it is possible to combine entertainment with an educational theme. This idea was taken up by Indian television in 1984; a soap opera called *We People* that emphasized family planning attracted top audience ratings. Now other nations are adapting the idea of combining entertainment with development.

We conclude that communication technology cannot alone lead to development, but under many circumstances it can be an important component in a development program.

REPORT

Communication Technology in Developing Countries

Heather E. Hudson
The University of Texas at Austin

Developing countries are characterized not only by poverty but also by diversity. In the poorest countries such as Bangladesh and Ethiopia, which rely on subsistence agriculture, per capita income is only about $200 per year. In the middle income countries such as the Philippines, Malaysia, and Brazil, which have a more mixed economic base, per capita income exceeds $1200. Although there has been dramatic urban growth in most countries, developing country populations are still overwhelmingly rural.

Of course, income is not evenly distributed throughout the populations, nor are services, including health care and education. Similarly, access to the basic infrastructure including roads, electricity, and telecommunications is likely to be very limited, especially in rural areas. Although the transistor radio has diffused throughout the developing world in the last 20 years, basic two-way communications are nonexistent in many areas. In some parts of Africa, there is only one telephone for more than 10,000 inhabitants, and people may have to walk for days to reach it.

But why should improvement in communications technology be a priority for nations that face so many development problems? Information is central to the development process. A community health worker who does not know how to treat a dying child, a farmer whose crop is being attacked by insects he cannot control, a rural cooperative that wants to get the best price for its harvest, and a shopkeeper who must order spare parts for the village water pump all share a need for information.

For decades, communication planners and researchers have looked to the mass media as a means of improving education and disseminating development information, but until recently two-way communication such as basic telephone service was ignored as a development tool. Now two-way radios are used in many isolated parts of the world to communicate with villages and rural development projects. New technologies can bring reliable communication to virtually any place that needs it, no matter how isolated. For example, satellites can be used for communication between villages and for teleconferencing. It is no longer necessary to extend wires and microwave repeaters across mountains, jungles, and deserts to reach isolated communities. Small satellite-earth stations can be installed in villages and powered by photovoltaic panels. A public telephone in a government office, clinic, or shop lets people communicate with their families in other communities, and with the hospital, government offices, and shops in the city.

In more affluent villages, people may have telephones in their own homes. Interactive telecommunications can be an important tool for improving access to health care and other social services. For example, in Alaska, village health aides communicate daily via satellite with doctors in regional hospitals to get advice on the diagnosis and treatment of village patients. The medical communications system is designed as an audio conferencing network so that health aides can learn by listening to the doctor's advice to other aides, and education courses can also be offered. A similar network, called LEARN/Alaska, is used to enable students in villages to take college courses offered by instructors in Anchorage. Residents throughout Alaska can testify in government hearings and get information from the state government in Juneau over the Legislative Teleconferencing Network, also via satellite.

Two-way communication can also be important for reaching isolated students and helping farmers to improve their crops. The University of the South Pacific (based in Fiji) offers tutorials for correspondence students over NASA's ATS-1 satellite. Agricultural extension officers can also use the network to get expert advice for farmers from the university's agricultural college in Western Samoa.

A satellite can also serve as a lifeline in an emergency. The ATS-1 network, also known as PEACESAT, has been used to coordinate disaster relief in the Pacific after earthquakes, typhoons, and outbreaks of cholera and dengue fever. After the 1985 earthquake in Mexico City, another NASA satellite, ATS-3, was the only reliable means of communications for several days when the telephone lines were down.

Today a few developing countries including Mexico, Brazil, India, and Indonesia have their own domestic satellites. Some other developing countries share regional or international satellites for domestic use. But access to telecommunications in the developing world is very limited. Although there are more than 600 million telephones in the world, more than two-thirds of the world's population has no access to telephone service. New York City has more telephones than the entire continent of Africa.

Yet the telephone can be an important development tool—for getting help in emergencies, getting advice from a doctor, getting help with correspondence studies, ordering supplies and spare parts, and finding the best price for village produce. With satellite technology, solid state components, and solar power the advantages of telecommunications can now be extended throughout the developing world.

Suggestion for Further Reading

Hudson, H. E. *The Telephone Comes to the Village.* Norwood, N.J.: Ablex, 1985.

REPORT

Planning Modern Communications Infrastructures

Herbert S. Dordick
Temple University

Communications infrastructures, both telecommunications and broadcasting, traditionally have been government monopolies in almost all countries. The United States has been the exception on both counts in that until recently telecommunications were a regulated private monopoly and broadcasting has always been and continues to be a commercial enterprise. Government control of broadcasting is believed necessary to protect the cultural boundary of a nation, to insure that the cultural values of a nation are preserved through the provision of broadcasting in the public service. Telecommunications entities, the nation's PT&Ts (Postal Telephone and Telegraph), have been operated as monopolies because they are perceived as natural monopolies. A natural monopoly exits when the cost of entry (i.e., of getting into business) is so high that the market cannot support more than one provider of the product or service. Another way of looking at this is that the provision of telecommunications products and services requires very large capital investments; and if more than one firm is in the marketplace, all will have to charge higher prices for telecommunications.

In the years following World War II, however, there have been important developments in information technology and radical changes in politics that have severely shaken the status quo. Chief among these has been the rapid development of the microprocessor and computer technologies and their introduction into broadcasting and telecommunications. As a result, the management and use of the electromagnetic spectrum has vastly improved. Broadcast channels can be more efficiently used, and higher frequencies have been opened so that new radio and television channels are available. Cable television, with its wide-band capability, essentially releases us from the tyranny of the spectrum by encasing broadcast signals in the coaxial transmission cable, thereby opening many more channels for the delivery of radio and television services. The twisted copper pair that provides POTS (Plain Old Telephone Services) can be more efficiently used through frequency and time division multiplexing, so that POTS are no longer plain.

Technology is not the only force that has altered the telecommunications broadcast environments throughout the world. Industry has recognized that telecommunications is an important strategic resource for growth: Information is a valuable commodity on domestic and world markets and requires telecommunications networks for its delivery. Satellites do not recognize national boundaries: Their "footprints" (broadcast coverage) spill over into neighboring countries. Citizens want more television programming; and if their nation cannot serve them, they purchase video cassettes and install satellite receiver antennas to pick up a neighboring country's broadcasting. Many countries have recognized that the only way to protect their cultural boundaries in the face of the satellite and the VCR is to provide more television programming by increasing the number of over-the-air transmissions or investing in cable television, or both. To do so requires large financial and human investments, especially for programming, and it is politically difficult, if not impossible, to raise the necessary revenue from taxes or receiver license fees. Reluctantly, France, the UK, New Zealand, and other countries have allowed advertising on their public service channels, are licensing privately owned commercial television broadcasting and cable television systems, and are stimulating local investment in film production through tax benefits.

Just as new broadcast technology has created new opportunities for broadcasters, new telecommunications technology has greatly reduced the cost of entry into the telecommunications markets. No longer need the PT&T or private monopoly be the sole provider of long-distance telecommunications. The local distribution loop can deliver a variety of enhanced communications services—high-speed data transmission, facsimile, and slow-scan video—thereby raising the possibility of multiple providers of new telephone services. Furthermore, the varieties of terminals that users want to connect to the telephone network have become as varied as the varieties of human communications styles, and it is virtually impossible, and indeed may be undesirable, for the PT&T to limit their use on the network as long as they satisfy reasonable technical standards. The traditional assumption that telecommunications is a natural monopoly is being challenged and with it, the monopoly status of telecommunications product and service providers, whether government or private.

Nations planning their broadcast and telecommunications structures are faced with many choices created by technology and the pressures of increasingly competitive international markets. Telecommunications and broadcasting are information resources, and as nations seek to enter the information economy, these resources are critical to economic growth.

Yet telecommunications and broadcasting provide important social services, and many countries believe that these services should be universally available at costs that are affordable for their citizens. Investments in new communications technology must provide for their efficient use as well as equitable distribution. Subsidizing one service by another, for example, subsidizing local telephone services by long-distance and business services, was the policy in the United States and continues to be so in many countries. However, this may lead to tariffs that discourage the rapid dissemination of the enhanced services that businesses need to compete effectively. And without cross subsidies, local rates may increase so that some people will no longer be able to afford a telephone.

Even in the highly industrialized nations, government funding of research has been important to the development of the information technology industries. In the United States, for example, the space program served as the catalyst for communications satellites, and the defense program has had a powerful impact on the nation's computer industry. Funds for information technology development must compete with other national needs: health and welfare services, education, transportation, agriculture, and defense, among others. Often, nations must import information technology at the cost of valuable foreign exchange. Financing information technology development and the construction of broadcast and telecommunications infrastructures must take into consideration both private and government investment. Consequently, governments must be concerned with regulatory policies and practices.

Regulatory options include the traditional government-owned monopoly, such as the PT&Ts of France, West Germany, and most of the industrializing nations; regulated monopoly, under which AT&T operated and towards which Japan appears to be moving; regulated competition, which is essentially the policy emerging in the United Kingdom; and the competitive marketplace, which the United States seeks to achieve. It is highly unlikely, however, that telecommunications and broadcasting can operate as free-market businesses. The electromagnetic spectrum is a limited natural resource, and cable television is likely to remain a local monopoly. Therefore, some form of government regulation is necessary to protect the public interest. Telecommunications systems are vast networks that must operate as integrated systems to function efficiently. Providers of network services must coordinate their activities to insure efficient system operation and government intervention may be necessary to insure that the public interest is served.

References

Dordick, H. S., and D. Neubauer. "Information as Currency: Organizational Restructuring Under the Impact of the Information Revolution." Forthcoming issue of the *Keio Communications Review,* Tokyo 1985.

Head, S. W. *World Broadcasting Systems: A Comparative Analysis.* Ch. 5. Belmont, Calif.: Wadsworth, 1985.

Homet, R. S. *Politics, Culture and Communications: European vs. American.* New York: Praeger, 1979.

Topics for Research or Discussion

▬▬▬▬ The Rogers and Larsen book *Silicon Valley Fever* describes both the highly successful entrepreneurship in the computer and semiconductor industries but also their negative side in terms of pollution, exploitation, and cultural impact. Review this book, then prepare a report or discussion of your evaluation of this example of high tech society. Include in your coverage the consideration of whether the Silicon Valley phenomenon can be transferred to other parts of the United State or to other countries as they seek to bolster their economic growth.

■■■■■■■ Third world countries complain that the wealthy countries, which comprise about 10% of the world population, monopolize 90% of the broadcast spectrum. This has been a heated issue at meetings of the World Administrative Radio Conference (WARC). Go to your library and do some research on WARC. Examine this issue and take a position on it. What are alternative solutions to the problem?

■■■■■■■ Often when an underdeveloped country installs a television broadcasting system, it finds that program development is so expensive that it is forced to import programming. By far the least expensive programs on the international market are American reruns of prime-time programs, some of questionable taste in a foreign market. Many countries have been concerned that the importation of such programs has negative cultural effects, sometimes calling this "media exploitation" or "media imperialism." They have therefore barred or limited importation. How do you evaluate this situation? Prepare a report on the topic and recommend courses of action. (TIP: Start by reading Katz et al., *Broadcasting in the Third World.*)

■■■■■■■ Communications technology has become an important topic to research in urban and land planning. Review the volume edited by Castells (reference list), then prepare a brief paper or discussion on the technology issues most relevant to this area of research. What types of joint research or theorizing might communications and land planning scholars undertake?

■■■■■■■ The free flow of information has been a controversial international topic, especially in debates in the United Nations Education, Scientific and Cultural Organization (UNESCO). In the broadest sense, it is meant to safeguard the developing countries of the world from media and information exploitation by the more developed countries. Examine this topic; then take a position on it. (TIP: Examine the index of the *Journal of Communication* for articles on the topic; also consult the Mankekar, Masmoudi, and Schiller volumes listed among the references to this chapter.)

References and Further Readings

Castells, M., ed. *High Technology, Space and Society.* Beverly Hills, Calif.: Sage, 1985.

Cherry, C. *World Communication: Threat or Promise?* New York: John Wiley, 1971.

Dordick, H.S. "Information Inequality." *Computerworld* (April 21, 1980):1 ff.

Edelstein, A., J. Bowes, and S. Harsel. *Information Societies: Comparing the Japanese and American Experiences.* Seattle, Wash.: University of Washington Press, 1978.

Fett, J. "Situational Factors and Peasants' Search for Market Information." *Journalism Quarterly* 53(1975):429–435.

Hardy, A. "The Role of the Telephone in Economic Development." *Telecommunications Policy* 4(1980):278–286.

Head, S.W. *World Broadcasting Systems: A Comparative Analysis.* Belmont, Calif.: Wadsworth, 1985.

Hudson, H.E. *The Telephone Comes to the Village.* Norwood, N.J.: Ablex, 1985.

Katz, E. and G. Wedell. *Broadcasting in the Third World: Promise and Performance.* Cambridge, Mass.: Harvard University Press, 1977.

King, J., and K. Kraemer. "Cost as a Social Impact of Information Technology." In *Telecommunications and Productivity,* edited by M. Moss. Reading, Mass.: Addison-Wesley, 1981, 93–130.

Kraemer, K.L., W.H. Dutton, and A. Northrop. *The Management of Information Systems.* New York: Columbia University Press, 1981.

Mankekar, D.R. *One Way Free Flow: Neo-Colonialism Via News Media.* New Delhi: Clarion Books, 1978.

Masmoudi, M. "The New World Information Order." *Journal of Communication* 24(1979):172–185.

Meyer, T., and A. Hexamer. "The Use and Abuse of Media Effects Research in the Development of Telecommunications Social Policy." In *Telecommunications Policy Handbook,* edited by J. Schement, F. Gutierrez, and M. Sirbu., Jr. New York: Praeger, 1982, 222–235.

Parker, E. "Social Implications of Computer/Telecoms Systems." *Telecommunications Policy* 1(1976):3–20.

Parker, E. "Communication Satellites for Rural Development." *Telecommunications Policy* 2(1978):309–315.

Parker, E., and D. Dunn. "Information Technology: Its Social Potential." *Science* 176(1972):1392–1399.

Parker, E., and H. Hudson. "Telecommunication Planning for Rural Development." *IEEE Transactions on Communications* 23(1975):1177–1185.

Picot, A., H. Klingenberg, and H.P. Kranzle. "Organizational Communication Between Technological Development and Socio-Economic Needs: Report from Field Studies in Germany." Vol. 6, *Communication Yearbook,* edited by M. Burgoon. Beverly Hills, Calif.: Sage, 1982, 674–693.

Pool, I. de Sola. *Technologies of Freedom*. Cambridge, Mass.: Harvard University Press, 1983.

Rice, R., and E. Parker. "Telecommunications Alternatives for Developing Countries." *Journal of Communication* 29(1979):125–136.

Rogers, E.M. "Communication and Development: The Passing of the Dominant Paradigm." *Communication Research* 3(1976):213–240.

Rogers, E.M. *Diffusion of Innovations*. 3d ed. New York: Free Press, 1983.

Rogers, E.M., and J. Larsen. *Silicon Valley Fever*. New York: Basic Books, 1984.

Rogers, E.M., and A. Picot. "The Impacts of New Communication Technologies." In *The Media Revolution in America and in Western Europe*, edited by E.M. Rogers and F. Balle. Norwood, N. J.: Ablex, 1983.

Schiller, H. *Who Knows? Information in the Age of the Fortune 500*. Norwood, N. J.: Ablex, 1982.

Schramm, W. *Mass Media and National Development*. Stanford, Calif.: Stanford University Press, 1964.

Thorngren, B. "Silent Actors: Communication Networks for Development." In *The Social Impact of the Telephone*, edited by I. de Sola Pool. Cambridge, Mass.: MIT Press, 1977, 374–385.

Tyler, M. "Telecommunications and Productivity: The Need and the Opportunity." In *Telecommunications and Productivity*, edited by M. Moss. Reading, Mass.: Addison-Wesley, 1981.

Tyler, M., M. Katsoulis, and A. Cook. "Telecommunications and Energy Policy." *Telecommunications Policy* 1(1976):21–32.

IV

NEW PERSPECTIVES

New media technologies necessitate that we update and revise our traditional theories of human communication. The two concluding chapters represent examples of this challenge. Chapter 13 explores the application of uses and gratifications theory to new technologies. Chapter 14 is a consideration of how computing and telecommunications technologies change the basic nature of our environment.

Extensions of Gratifications Theory

Frederick Williams
Amy F. Phillips
Patricia Lum

As well-stated by gratifications theorist Elihu Katz (et al., 1976), why individuals use a communications medium is a much more theoretically valuable question than questions pertaining to details of the medium or particular instances of use. In the present chapter, we explore extensions of uses and gratifications theory as it may apply to the study of new media.

TOPICAL OUTLINE

Going Beyond Mass Communication

Approaches and Implications
Functionalist Approaches
Structuralist Implications
Action/Motivation Implications

Applications to Media Technologies
Overview
Cable TV
Video Cassettes

This chapter appeared in an earlier form under the title "Gratifications Association with New Communication Technologies," in K. E. Rosengren, L. Wenner and P. Palmgreen, Eds. *Media Gratifications Research.* © 1985 Sage Publications, Inc. It is used by permission of the Sage Publications, Inc. and the authors.

Toward a Framework for Studying New Technologies: Six Topics for Research

GOING BEYOND MASS COMMUNICATION

Uses and gratifications theory has historically been applied to mass media, but it has always held promise for the study of other media as well, including the so-called new technologies. Certainly, the proliferation of new communication technologies may affect the structure of communication in society and make available a greater range of choice for satisfying communication needs. New media uses may complement uses already studied. Previously identified uses may shift to new media from old ones, providing fresh insights into the relationship between media use and gratifications. Further, uses and gratifications theory incorporates concepts that provide a base for developing a framework for research into the adoption of new technologies.

A challenge, as posed by Palmgreen (1984), is for researchers to "adapt and mold the current conceptual framework to deal with new communication technologies." In essence, the likely questions raised by this challenge are the topic of this section of the present chapter.

To the user, most communication technologies are mainly extensions of existing media, for example, in the sense that video cassettes or disks are related to broadcast television, or the new electronic text services are extensions of traditional print. Important, and especially relevant to the present chapter, is that the new media offer an increased number of alternatives for both access to and interaction with message stimuli. Some of the technologically based contributions include

- Making distance all but irrelevant (communications satellite)
- Freeing television from the restrictions of broadcast schedules (video tape)
- Providing for nonlinear access to information (computers and video disks)
- Offering nearly unlimited availability of two-way voice or text communications (mobile telephone and computer teleconferencing)

- Transporting many simultaneous message or program choices (coaxial cable and fiber optics)
- Bypassing the printing and transportation requirements for the transmission of textual information (video and teletext)

These new opportunities for access have more generalized psychological consequences for the human communicator. As already mentioned, there are more choices and hence more alternatives for gratifications (as in having 30 rather than five channels of television or in the added alternatives of disks, tapes, cable, and pay-TV channels). Mobility is greatly increased thereby opening new opportunities to engage in gratifications-seeking independent of locational restrictions (as in viewing movies at home rather than at a local theater). There is increased freedom to create personal schedules, as reflected in video tape use or in working at home during hours of one's choice. Finally, there is the increased opportunity for interactivity with media (as with random access video disk or computers) or via telecommunications networks (as with teleconferencing).

Presumably, it is within these more psychological ramifications of the new media, rather than their technical aspects, that we can pose interesting applications or implications of uses and gratifications theories.

APPROACHES AND IMPLICATIONS

Functionalist Approaches

Our brief review of relevant uses and gratifications studies is organized along the lines of the three-category distinction made by McQuail and Gurevitch (1974) among functional, structural-cultural, and action-motivation approaches to the topic.

From the broad perspective of drive-reduction theory, all uses and gratifications studies are functionalist in the sense that they look upon the audience as being active gratifications-seekers in interaction with media rather than passive receivers. However, in our view, most uses and gratifications research (the "voluntaristic and selective nature of the interaction between audience and mass media"—Levy and Windahl, 1984) has served to limit this notion of interaction. For example, the active audience has been defined as the "before- ," during- ," and "after-exposure" audience, with activity being either present or not present. Blumler (1979) suggests that audience activity be treated as a variable and that activity be operationalized according to the foregoing temporal sequence.

Although the preceding idea is not unique to uses and gratifications theory, it does remind us of the process nature of communication and that gratifications at one time may give way to others later in the media-use process. All such thinking raises questions as to whether a user's increased control over a medium—as many new technologies provide—gives additional importance to the before-during-after sequence. Moreover, is there gratification that might be directly associated with the feeling of control over a medium?

Levy and Windahl (1984) extend the conceptualization of audience activity as a variable construct so that it might be applied not only to content choice but to the choice of which medium to use. They base their study of TV news viewing in Sweden on the theoretical assumption that audiences exhibit varying types and amounts of activity in different communication settings and at different times in the communication sequence. They propose a typology of audience activities based on two dimensions. The qualitative orientation of audience includes the three nominal values of (1) audience selectivity, (2) audience involvement, and (3) audience use (psychological or social utility of media) along a temporal dimension including activities before, during, and after exposure. Different types and degrees of activity may be associated with particular phases of the communication process. No direct links were found between activity before, during, and after exposure. But their study did find that levels of audience activity covary directly with gratifications sought and/or obtained and suggests that audience activity and gratifications "stand as important intervening variables in the communication process" (1984, p. 74).

The latter raises the direct question of how the expectations of gratifications can be substantially increased by the presence of additional media, content, or operation (as with interactivity) alternatives. For example, what if an interactive videotext system is added to one's sources of current news? Would it change the qualitative orientation? If such a system allowed the user to pursue selected news topics in depth, would audience activity in news-seeking become much more reflective of personal rather than media agenda-setting?

Structuralist Implications

What ideas garnered from the structuralist perspective can be helpful in evaluating new technologies? The media structure in a person's environment is important in determining the path taken to gratify a communication need. Although ostensibly not using a structural perspective, Blumler (1979) was able to distinguish four factors for differentiating gratifications associated with news about political campaigns (surveillance, curiosity, diversion, and personal identity) in Great Britain. However, Becker (1979), in an attempt to do a broader factor analysis of gratifications associated with political information sought and received in Syracuse, New York, across several campaigns, did not find such a clear factor structure. One possible explanation for the contrasting results may lie in the structural differences in the ownership, operation, programming, or content of newspapers and television in Great Britain as compared to the United States.

Another possibility, arising from the acknowledgment of structural differences, but extended to the realm of individual choice-making, is that people make tradeoffs between various media according to what is available and accessible to serve a particular perceived need. Blumler (1979) notes that housewives without a telephone show a significantly greater use of television for serving a personal identity function than would otherwise be served by social contact through use of the telephone.

Certainly, then, from the broader environmental view, there is the question of how or whether new technologies will change environmental alternatives for media gratifications. Already, annual Nielsen summaries are showing some tradeoff in the United States between audiences' increased uses of tape and cable TV rather than traditional network broadcasting for their entertainment needs. Further, tape rentals are visibly cutting into traditional movie-going (a condition reminiscent of the effect of television's introduction in this country). Indeed, we are seeing modifications in the choice environment. Some of these may be short-lived, like video games, and others may forever change the media environment (as television did and as video cassettes now seem likely to have done).

Another example of environmental change, although seldom addressed in uses and gratifications research, is in the effects of new communications technologies on organizational environments. Management information systems, word processing, electronic mail, and other new technologies of office automation are visibly changing the nature of organizational communications and, presumably, the associated uses and gratifications as well.

Action/Motivation Implications

How do we account in detail for the dynamics of media and content selection? Van Leuven (1984), for example, presents a two-level expectancy theory "capable of handling media and message selection processes at once" (p. 426). This type of theory, originally suggested by McQuail and Gurevitch (1974), is an action/motivation theory focusing on individual users, their choices of media behavior, and, perhaps in response to Swanson's (1977) criticism of lack of attention paid to perception of meaning attached to media and messages, on the meanings and expectations they attach to those choices.

Other authors have proposed expectancy models (Galloway and Meek, 1981; Palmgreen and Rayburn, 1982), but all are similar in that they view either behavior, behavioral intentions, or attitudes as a function of (1) expectancy—the perception of an object's possession of a particular attribute or that a certain behavior will lead to certain consequences—and (2) evaluation—that is, the degree of "effect, positive or negative, toward an attribute or behavioral outcome" (Palmgreen, 1984). Palmgreen and Rayburn's (1982) process model holds that the products of beliefs and evaluations influence the seeking of gratification, which influences media choice and consumption. This then results in the perception of certain gratifications obtained, which then feeds back to reinforce or alter an individual's perceptions of the "gratification-related attributes of a particular" medium, message (content), or program genre.

To date, we know very little of the beliefs and evaluations associated with new media alternatives. Even if we do learn about beliefs, are they predictive of behavior? This suggests the need for longitudinal studies of media use and the varying stages of activity and decision-making processes. For example, Atwood and Dervin (1981) have based several studies on the notion that as people move through time and space they encounter gaps or inconsistencies and try to use information for assistance in reducing uncertainty. Put another way, the researchers explain that information-seeking behavior can best be analyzed by examining how people view their constantly changing environments and how their information needs correspond with those changes. Furthermore, information-seeking is construed as a highly active process; messages are merely words or data until they are actively processed by an individual and constructed as pieces of information. How, then, does such thinking apply to new technologies? Will increased availability of information be matched with increased use? Or if new media are made available for the public good (as with libraries), what kind of training or motivation will be required so as to promote use?

It is possible, too, to conceive of broad dimensions of media attitudes such as found in Phillips's (1982) pilot research with university students. Multivariate scaling techniques clearly indicated that the students were differentiating communications alternatives such as video cassettes, cable TV, or even computers along rather basic attitudinal dimensions. This raises questions not only about how such attitudes mediate a person's expectations of satisfaction from using a particular medium but also about how those attitudes become socialized in the first place.

It has also shown that content may be perceived not just as information but as other forms of stimulation as well. For example, social presence theory (Short et al., 1978) suggests that individuals associate different degrees of personal contact with various media, and that choice of a medium may not be for purposes of informational content but for the affective consequences. This holds particular implications for those new technologies (especially computers) that are felt by some to be markedly impersonal as media. How can the social presence of a medium be enhanced? Presumably this is as much a question of rhetorical choices and capabilities as it is one of technology. A telephone call, although lacking in visual cues and suppressing some paralinguistic cues, can nevertheless be highly personal if the appropriate language is used. The same can be applied to computer-based teleconferencing.

In all, we need more insights into the details of human and media interaction before we will be able to fully comprehend how such interaction contributes to need fulfillment. This includes examining the spectrum of highly interactive to not-very-interactive technologies in the gamut of media alternatives. In some instances, traditional distinctions among media may be blurring, thus further complicating the research picture. For example, how will we account for the new interactive or two-way media (e.g., asynchronous conferencing via computer links) that blur traditional distinctions between sender and receiver? Or does not the greater availability of program alternatives offered by cable, disk, tape, and eventually, direct broadcast satellites begin to blur the typical reference to television as a mass medium? In short, even if we do succeed in relating media attitudes to the dynamics of gratifications, it is likely that for the new media still finding their way into use, we will also have to incorporate models of attitude change.

APPLICATIONS TO MEDIA TECHNOLOGIES

Overview

Few uses and gratifications researchers have focused on studying new communications technologies. This may be because the most pressing research issues have had less to do with certain types of media than with conceptualizing the behavioral and social processes of selection and gratification. Although uses and gratifications approaches basically provide descriptive information about media use, media characteristics have changed and are continuing to change over time. One reason for studying the new technologies is to examine how, or if, gratifications change with media characteristics. Another is to gain further understanding of how new media are perceived and used.

Research into the new technologies falls primarily into the realm of function or utility studies, drawing inferences from uses rather than actually measuring gratifications. These studies are of value, since it is desirable to view uses and gratifications not of one medium alone but within the context of a person's total media environment. Communication needs may be better defined through examining changing human-and-media interactions. In the following few pages, several studies dealing with the uses of new technologies are reviewed. Additionally, some observations on the telephone—an old medium but with many of the characteristics of newer technologies—will also be considered.

Cable TV

Although cable television has had a long history of simply extending the signals of over-the-air broadcasters, the more modern and ambitious vision is one of wide selection and broadband communications services, including interactive ones. These can range all the way from televised entertainment to interactive banking.

One study that deals directly with expectations associated with cable television points out the requirements for researchers to expand their concept of what types of specific gratifications these may include. Shaver (1983) conducted focus group interviews and found dimensions of functions relating to the unique aspects of the medium. The two most frequently mentioned motives for viewing cable TV were the variety provided by the increased number of channels and programming choices and the control over viewing associated with the flexibility of programming. Although the study does not group dimensions in this manner, the motives fall into content (religious programs), structural (variety), service (better reception), and psychological needs (companionship) categories. Some of these gratifications have also been associated with traditional broadcast television (such as general surveillance), but others are relatively linked to the cable TV medium.

In a study conducted for the NCTA (Opinion Research Corporation, 1983) people who chose whether or not to subscribe to cable TV were loosely classified into a series of five user groups. Although better reception was still mentioned as the primary reason for subscribing to cable, certain other distinctions between users were defined. Three user groups reflected the following characteristics:

1. Undifferentiated users of the medium, who will watch everything and are early adopters of cable.
2. The "entertain me" group, who primarily seek entertainment and diversion and are likely to be pay-TV subscribers.
3. The "Basic but . . ." group, who are more differentiating in their use. Specifically, they want sophisticated, intellectually stimulating, children's, family-oriented, or information programming, and are likely to seek out home services and other special services.

The foregoing study shows that functions relating both to the form and to the content of the medium are associated with a technology. Furthermore, a distinction seems to be evident between people who seek out entertainment or escape and those who seek a variety of specific services, a point we raise later in consideration of the new text services.

Video Cassettes

Video cassettes differ from television as audio cassettes do from radio. Although there are similarities in content and style, the medium itself allows the audience choice in the temporal dimension and perhaps along the qualitative dimension proposed by Levy and Windahl (1984). Specifically, video cassette recorders, which are rapidly growing in popularity and use, offer the audience the opportunity for time-shift viewing (Waterman, 1984). Unlike video disks, which for the present can only be obtained with prerecorded content, blank video cassettes can be used for recording broadcast or cable TV programs for viewing at a later time.

Studies (e.g., university students studied by Phillips, 1982) consistently indicate that time-shifting and prerecorded theatrical films (e.g., Waterman, 1984) are the most popular uses of video cassette machines. Further uses include playback of home-produced tapes, self-help programs (e.g., exercises), educational materials, music videos, and uses coupled with a camera. (See also studies by Levy and Fink and by Rue as discussed in Palmgreen, 1984.)

The foregoing uses of video tapes reflect especially on the earlier mentioned psychological consequences of choice, time, and mobility associated with certain new media. The program content may often be the same as in the older medium (e.g., television or films), but the additional advantage is now related to circumstances of use. Nor should we overlook the fact that cost is often reduced, as with the case of viewing films. In all, the implications are that a new media technology may offer less in the way of gratifications itself but make traditional gratifications (e.g., TV shows and movies) more easily obtainable.

Cassettes also hold implications for the structure of the film and television program industry, thus contributing to media environmental change. One advantage in the production and distribution of video cassettes is that new productions can be economically viable without having to appeal to large audiences. Thus we may see the proliferation of highly specialized cassette programs along with those geared to the mass market.

Interactive Services

Of course, our most prominent interactive media technology, the telephone, which has been with us for over a century, embodies many of the promises of the new media. Other interactive technologies of current interest include two-way cable (e.g., the Warner-Amex Qube system) and various forms of text services, both broadcast or cable transmitted (teletext) or available via phone networks (videotext). The interactive capability of such services is used for various ends, ranging from the selection of programs to a full two-way message system. Psychological consequences include greater selection, more personal control over selection, and the sense that one can be a communications source as well as receiver.

It is becoming commonly known that interactive television services such as Qube are not as popular as was originally envisaged. A reason frequently given is that the public has grown to expect mainly entertainment and escape—that television offers respite rather than involvement. Another reason is that for interactive services to hold longer range customers, they must offer advantages in time, convenience, or cost. It is noteworthy that while there has been little growth of interactive television on a mass media scale, there has been steady growth of specialized information services such as those offered by Dow Jones News Retrieval, The Source, or CompuServe.

Our existing theories of uses and gratifications are already a basis for explaining why expectations associated with entertainment (relaxation and escape) are not specifically fulfilled by interactive media. Also, it may be mainly convenience that explains specific gratifications associated with the information services. Deeper needs and their gratifications may have little to do with new understandings of the latter services.

However, one study of interactive services did bear upon functional dimensions of use. Dozier and Ledingham (1983) identified two key dimensions in use of interactive banking, shopping, and home security in San Diego; these were surveillance and transaction. The surveillance mode, a read-only status, was found to be more attractive to users than the transaction, read-write, interactive mode. Although not based on formal evidence, this finding prompted the researchers to conclude that, all other factors being equal, the surveillance function of this information utility will be adopted more rapidly than the transaction function.

The information retrieval systems are sometimes associated with a depersonalization factor reflective of the social presence studies (Short et al., 1976). Take, for example, systems used mainly for electronic banking, computerized shopping, and home security. The same Dozier and Ledingham study revealed that while there are perceived advantages such as time savings and convenience, such services are often viewed as not being worth the disadvantage of the loss of social interaction. Thus, while the desire for convenience is served, the need for getting out of the house and engaging in daily social interactions is not met for some users.

The Telephone

We have chosen to give the telephone some attention in this discussion because for years it has provided interactive services that are highly accepted in all modern societies. Ironically, however, the telephone has received very little attention from social scientists and almost none in the uses and gratifications literature. Its study in retrospect, as well as in current changes, can contribute to our understanding of uses of new technologies.

Keller (1977) has postulated that telephone use has two dimensions, the instrumental and the intrinsic (i.e., talking on the telephone for its own sake). The concept of the telephone's intrinsic uses was further borne out in Wurtzel and Turner's (1977) study of New York City residents whose telephone service was temporarily cut off. Contrary to the expectation that other forms of communication would be substituted for the telephone, it was found that certain uses of the telephone, particularly for making social calls, were not substitutable.

An ethnographic study of telephone uses (Phillips, Lum, and Lawrence, 1983) revealed that people tend to differentiate between business and pleasure calls. Business-related calls and some types of social calls to business associates in various relationships to the caller are thought to be more appropriate for the workplace, while social calls and some business calls are more appropriate for the home. Thus use is associated with location and relationship along the business-pleasure dimension.

Lum (1984), in conducting a series of in-depth interviews about phone use by senior citizens in Hawaii, found two major perceived-needs dimensions satisfied by telephone use. The convenience dimension was divided into social calls and informational calls, which includes the need to "get things done" and to "get information to get things done." The second category was that of contact calls, which included social-obligation calls (perceived as necessary to uphold the notion of one's social role), "I care," or extremely affective calls meant to "brighten one's day," say "I love you," or to lift morale.

Another important observation was that respondents did not consider their telephone use similar to the singular use of a communications medium. Instead, there was far more emphasis upon its content and use as an interdependent part of everyday life. Lum found that other media were brought up as being important, especially when respondents were asked to imagine what would happen if their phone were taken away. For instance, one person mentioned that without the telephone the world would "go slower" since there would be no direct contact; television would be watched but could not substitute for telephone use.

Teleconferencing

Uses of technologies have various forms in conferencing applications. These can include audio, video, facsimile, data, or text links, as well as combinations of any of them. Most studies of teleconferencing have centered upon factors of effectiveness rather than upon needs satisfaction.

Although audio conferencing has long been available through public telephone services, its use has been surprisingly modest, given all the attention on modern teleconferencing. Many studies (e.g., Dutton, Fulk, and Steinfield, 1982) have shown that with proper management video conferencing can be effective in meeting situation-specific communication needs of corporate communication.

Computer teleconferencing has grown in recent years, especially in organizational environments where messages may be stored for forwarding—as in asynchronous conferencing. In fact, the computer may be losing its forbidding, impersonal image as more individuals gain experience in computer-mediated communications. Hiltz and Turoff (1979) suggest that there is an affective dimension present in both asynchronous and synchronous computer teleconferencing. People may anthropomorphize and use computers in roles ranging from psychiatrists to "partners in crime."

Computer-aided networks, such as electronic bulletin boards and services like The Source, are becoming more accessible as more people purchase home computers. The choice of using such systems seems to reflect various motives. In addition to accomplishing specific tasks (as discussed with information services), these motives include the desire to establish and maintain contacts outside one's own geographic areas, the need for access to schedules, and the like. Hiemstra (1982) and Rice (1980) suggest that such uses do not always depend on interpersonal involvement, hence, the technical restrictions of the medium do not so much impede effectiveness.

Yet Hiltz (1978) and Phillips (1983) found that people are very active in their socioemotional use of computers for working out problems and in using face-saving tactics or stream of consciousness thinking, even when the conference they are involved in is task-oriented. It may be inferred from these and other studies emerging in the literature that there are many different types of needs and gratifications associated with computer-mediated communication that have not yet been assessed.

Apparently, a personal utility factor is also involved. Turkel (1984) points out that familiarity with "computerese" is important since computer-related terms and concepts now permeate our lives.

Electronic Mail

Some of the computer-based conferencing systems have the same characteristics as electronic mail. Messages are sent to a central computer where they can be forwarded at the receiver's request. For several years, the U.S. Postal Service offered an ECOM service whereby computer-originated mail was electronically distributed to the post offices nearest the addressees, then printed and mailed first class. Although that service has now been discontinued, various forms of it are offered by Western Union, Federal Express, and MCI. To the authors' knowledge there are no major social scientific studies of electronic mail of this type. Yet, the implications are that being able to send mail nearly instantly from a home computer keyboard must have some visible gratifications, or the electronic mail business will have no commercial value. Are factors of choice and mobility, or just simple convenience, important here? Or are there other, deeper satisfactions such as the ease of keeping in touch with colleagues or friends? Is electronic mail only perceived as a convenient substitute for traditional forms of postal services, or does it offer new satisfactions?

Computer networks within organizations sometimes offer messaging systems also referred to as electronic mail. In a study of one such system, Steinfield (1983) found that the mail system was used for a wide variety of purposes, some that bore little resemblance to the standard teleconferencing-based uses. A search for major dimensions of use uncovered the task and socioemotional use clusters. The former were self-evident, mostly the act of transferring or obtaining information. The latter, however, might appeal to uses and gratifications theorists. It related to the maintaining of personal relationships, feeling a part of the organization, and being in touch. Social uses were reported almost as frequently as were task-oriented ones. Again, personalization is an important gratification associated with some media technologies.

TOWARD A FRAMEWORK FOR STUDYING NEW TECHNOLOGIES: SIX TOPICS FOR RESEARCH

What then should be considered in the application of uses and gratifications approaches to the study of the new media technologies? We see six topics of special concern that are beginning points in the development of a framework for research. These include expanded choice, interactivity, personalness, new types of gratifications, audience concepts, and a broadened theoretical focus.

1. Expanded Choice

Although we initially took the position that the new media are less new than they are extensions of traditional media, many new circumstances of use contribute to a wide array of new or modified uses and gratifications. To understand how these uses fit together and how media are used to satisfy needs, we must look to the individual's total media environment. Changes might be seen as initially structural; for example, there are more alternatives from which to choose. But changes can also be seen in terms of specific modifications of choice, as, for example, in how the introduction of pay cable services has altered the uses of other media. As a result, media use may become more highly differentiated in serving communication needs. To continue the example, some uses of cable may substitute for uses previously assigned to broadcast television, but new uses will also appear that are complementary and not substitutes for old media uses. One new use might be to purchase special informational programs on the stock market.

2. Special Qualities of Interactivity

The literature frequently suggests the importance of the emerging contrast between technologies that are interactive and those that are noninteractive. Surely, they may serve different communication needs. This contrast shows up particularly in experiences with the new text services. On one hand, services have been designed to meet a wide variety of media needs (e.g., traditional publishers have attempted to distribute electronic magazines), many of which presumably would be definable in terms of traditional uses and gratifications. On the other hand, there are the gratifications associated with specific, utilitarian services (e.g., banking) that are so specifically task-related that more general gratifications concepts seem irrelevant. For example, using videotext to make travel reservations is fundamentally different from deciding to view a movie with your video cassette player.

3. Personalness

Interactivity looms in importance with its potential for personalness, both in offering a wide variety of choice (as in an interactive pay cable system) and in the potential for interpersonal transactions between individuals or among groups of communicators (as in electronic mail). These are both qualities that especially distinguish new media from old. There are also the rhetorical implications of developing personalized communication styles for use in new media (as, for example, in teleconferencing).

4. More Specific and Personalized Gratifications

There is greater opportunity for unique, personalized gratifications, some distinguished from traditional media rewards. Text services present a potentially infinite range of alternatives for the consumer-user. As contrasted with a mass communications medium where uses and gratifications research has often associated certain satisfactions with a given medium, text services, by their nature, could offer any gratifications allowable by the nature of the medium (and, of course, viable in the marketplace). This leads media development away from products (especially, films and television shows) that must capture a large mass market to survive. There is a greater potential in text services for more esoteric materials to be made available. They need not have mass appeal, a characteristic that fits the current concept of demassification of media.

Interactive media such as computer mail or bulletin board services also allow people the opportunity to be message disseminators. As such, they may help gratify a need to circulate a message to many receivers simultaneously or asynchronously. Also, some of the most used text services are those in which the satisfied need reflects accomplishment of a specific task (e.g., booking theater tickets) rather than a traditional media-related satisfaction (e.g., being entertained). The task may have emotional dimensions as well. For example, people may use computer bulletin board services to meet new friends (even romance!) or to share ideas concerning a specific topic of interest.

5. The Concept of Audience

The vastly increased choice potential of certain new media (e.g., video cassettes and text services) changes the concept of the relation of media-and-audience vis-á-vis selection. If this is so, the concept of audience becomes problematic, and the direction for future research should look at individuals as participants. Perhaps many assumptions about audience need to be reexamined. For example, as the degree of media and content self-selection increases, audience takes on more the nature of individual participants than aggregate groups. General models for examining audience gratifications from traditional models become less useful.

6. Expansion of Theoretical Focus

Given the range of choice and the complementarity of various new media, many intervening variables become important. The different approaches to the study of uses and gratifications—e.g., the functional versus the motivational approach—used in previous research when applied to one medium become inadequate. New models that incorporate utilitarian functions, range of choice, the phenomenon of personalization of a medium, and the temporal dimension of attitude must be posited and tested in the framework of communication gratifications. This opens up the opportunity to further, not negate, the area of uses and gratifications research. Such new research must be prepared to map its own way. Those measures of gratifications previously derived were mostly limited to the concept of a mass media audience.

New dimensions need to be measured that look at the audience as participants. And as the communications choices can be defined in conceptual terms more closely related to basic human needs, perhaps we can map those needs more clearly than in the past upon existing social-psychological theories of drives and drive reduction behaviors. Only when uses and gratifications theory can be interpreted relative to the larger context of psychological theory will it contribute to our greater understanding of the human condition and the complex roles that communications serve therein.

Topics for Research or Discussion

━━━━━━ Uses and gratifications is not an uncontroversial theory. Some critics accuse it of "monocausality," meaning, among other things, that it oversimplifies the explanation of why we use certain media. Another criticism is that uses and gratifications theorists have never precisely anchored their theory in existing theories of motivation and behavior. Prepare a critical essay on uses and gratifications theory as applied to the new technologies. (TIP: Consult Rosengren et al., 1985, for a review of contemporary opinions on this topic.)

━━━━━━ If you have Cable Network News's *Headline News* in your community, analyze the content of this program, both before and after one of the major network news programs each evening. If a person viewed only one or the other of these sources of evening news, what types of differences in the knowledge of the news might result? What do you feel are the major contrasts in programming strategy? What type of future do you see for headline services?

━━━━━━ The *Social Psychology of Telecommunications* (New York: Wiley, 1976), by Short, Williams, and Christie, was one of the first major collections of research reports to examine attitudinal aspects of media technologies. Review this book, then prepare a brief report that addresses the following questions: What was the major contribution of this research? How could the research have been improved? What research should be done next, including that involving new media technologies?

━━━━━━ Select an example of a modern media technology and evaluate likely user behavior in terms of uses and gratifications. For example, how might the use of videotext be explained from this standpoint?

━━━━━━ Similar to the emphasis upon uses and gratifications in the present chapter, what other theories might you employ to interpret uses of communication technologies? Describe one such theoretical approach, then examine the relative advantages and disadvantages.

References and Further Readings

Atwood, R., and B. Dervin. "Challenges to Sociocultural Predictors of Information Seeking: A Test of Race Versus Situation Movement State." Vol. 5, *Communication Yearbook,* edited by M. Burgoon. New Brunswick, N.J.: Transaction Books, 1981, 549–569.

Becker, L. B. "Measurement of Gratifications." *Communication Research* 6(1979): 54–73.

Blumler, J., and E. Katz, eds. *The Uses of Mass Communication.* Beverly Hills, Calif.: Sage, 1974.

Blumler, J. G. "The Role of Theory in Uses and Gratifications Studies." *Communication Research* 6(1979): 9–36.

Dervin, B. "Mass Communicating: Changing Conceptions of the Audience." In *Public Communication Campaigns,* edited by R. Rice and W. Paisley. Beverly Hills, Calif.: Sage, 1981, 71–88.

Dozier, D. M., and J. A. Ledingham. "Perceived Attributes of Interactive Cable Services Among Potential Adopters." Paper presented to the Human Communication Technology Special Interest Group, International Communication Association Annual Convention, Boston, Mass., May 2, 1982.

Dutton, W. H., J. Fulk, and C. Steinfield. "Utilization of Video Conferencing." *Telecommunications Policy* (September 1982): 164–178.

Galloway, J. J., and F. L. Meek. "Audience Uses and Gratifications: An Expectancy Model." *Communications Research* 8(1981): 435–449.

Hiemstra, G. "Teleconferencing, Concern for Face, and Organizational Culture." Vol. 6, *Communication Yearbook,* edited by M. Burgoon. Beverly Hills, Calif.: Sage, 1982, 874–904.

Hiltz, S. R. "Controlled Experiments with Computerized Conferencing: Results of a Pilot Study." *Bulletin of the American Society for Information Science* 4(1978): 11–12.

Hiltz, S. R., and M. Turoff. *The Network Nation: Human Communication Via Computer.* Reading, Mass.: Addison-Wesley, 1978.

Katz, E., M. Gurevitch, and H. Haas. "On the Uses of the Mass Media for Important Things." *American Sociological Review* 38(1973): 164–181.

Keller, S. "The Telephone in New (and Old) Communities." In *The Social Impact of the Telephone,* edited by I. Pool. Cambridge, Mass.: MIT Press, 1976.

Kling, R. "Social Analyses of Computing: Theoretical Perspectives in Recent Empirical Research." *Computing Surveys* 12(1980): 61–110.

Levy, M. R., and S. Windahl. "Audience Activity and Gratifications: A Conceptual Clarification and Exploration." *Communication Research* 11(1984): 51–78.

Lometti, G., B. Reeves, and C. Bybee. "Investigating the Assumptions of Uses and Gratifications Research." *Communication Research* 4(1977): 321–338.

Lum, P. "Telephone Use By Senior Citizens: Community Snapshot." Unpublished paper. Annenberg School of Communications, University of Southern California, Los Angeles, Calif., May 1984.

McQuail, D., and M. Gurevitch. "Explaining Audience Behavior: Three Approaches Considered." In *The Uses of Mass Communication: Current Perspectives on Gratifications Research,* edited by J. Blumler and E. Katz. Beverly Hills, Calif.: Sage, 1974, 287–301.

Opinion Research Corporation. "Segmentation Study of the Urban/Suburban Cable Television Market." Prepared for the National Cable Television Association. Princeton, N.J.: Opinion Research Corporation, 1983.

Palmgreen, P. "The Uses and Gratifications Approach: A Theoretical Perspective." Vol. 8, *Communication Yearbook,* edited by R. Bostrom. Beverly Hills: Sage, 1984.

Palmgreen, P., and J. D. Rayburn. "Gratifications Sought and Media Exposure: An Expectancy Value Model." *Communication Research* 9(1982): 561–580.

Phillips, A. F. "Attitude Correlates of Selected Media Technologies: A Pilot Study." Unpublished paper. Annenberg School of Communications, University of Southern California, Los Angeles, Calif., 1982.

Phillips, A. F. "Computer Conferences: Success or Failure?" Vol. 7, *Communication Yearbook,* edited by R. Bostrom. Beverly Hills: Sage, 1983, 837–856.

Phillips, A. F., P. Lum, and D. Lawrence. "An Ethnographic Study of Telephone Use." Paper presented to the Fifth Annual Conference on Culture and Communication, Philadelphia, Penn., March 1983.

Rice, R. E. "The Impacts of Computer-Mediated Organizational and Interpersonal Communication." *Annual Review of Information Science and Technology* 15(1980): 221–249.

Rosengren, K. E., L. A. Wenner, and P. Palmgreen. *Media Gratifications Research.* Beverly Hills, Calif.: Sage, 1985.

Shaver, T. L. "The Uses of Cable TV." Unpublished Master's Thesis, University of Kentucky, 1983.

Short, J., E. Williams, and B. Christie. *The Social Psychology of Telecommunications.* New York: John Wiley, 1976.

Slack, J. "Surveying the Impacts of Communication Technologies." Vol. 5, *Progress in Communication Sciences,* edited by B. Dervin and M. Voigt. Norwood, N.J.: Ablex, 1984.

Smith, A. *Goodbye Gutenberg: The Newspaper Revolution of the 1980's.* New York: Oxford University Press, 1980.

Steinfield, C. S. "Communicating Via Electronic Mail: Patterns and Predictions of Use in Organizations." Dissertation, Annenberg School of Communications, University of Southern California, December 1983.

Turkel, S. *The Second Self.* New York: Simon & Schuster, 1984.

Van Leuven, J. "Expectancy Theory in Media and Message Selection." *Communication Research* 8(1984): 425–434.

Waterman, D. "The Prerecorded Home Video and the Distribution of Theatrical Feature Films." Paper presented at the Arden House Conference on Rivalry Among Video Media, Harriman, N.Y., April 1984.

Wurtzel, A. H., and C. Turner. "What Missing the Telephone Means." *Journal of Communication* 27(1977): 48–56.

14

Dimensions for New Research

Frederick Williams
Ronald E. Rice
Herbert S. Dordick

In the larger and longer-range views, communication and computing technologies are changing the nature of the human environment. This closing chapter explores likely dimensions of these impacts.

TOPICAL OUTLINE

A Concept of Behavioral Impacts

Six Dimensions of Impact

1. *Attitudes*
2. *Time and Space*
3. *Connectivity*
4. *Mobility*
5. *Increased Choice*
6. *Socialization*

This chapter appeared in an earlier form as "Behavioral Impacts in the Information Age," in Brent D. Ruben, ed., *Information and Behavior,* Volume I, copyright © 1985 by Transaction, Inc. Used by permission of Transaction, Inc. and the authors.

A CONCEPT OF BEHAVIORAL IMPACTS

Whereas the themes of the chapters in the preceding section of this book focused upon the details of technological applications in differing communication contexts, in this chapter we take a much broader view of the social impacts of communication technologies. In fact, in approaching the study of communication technology and behavior, one finds an increasingly recurring picture of different levels of analysis. On the one hand, there are the details of the technologies and their interface with the human user. At a particularly detailed level, this might be called human-factors research, in which such questions are found as follows: What is the optimum way to display numeric information? What is the most effective layout for a computer keyboard? How can help menus be best incorporated into computer programs? At what distance are people most comfortable watching television?

At a slightly more general level, we inquire into how the technologies can be integrated with our behaviors, for example, in such questions as: How can we most effectively write with a word processor? What types of video cassettes are most attractive for relaxation or entertainment? What types of choices do we make when deciding whether to make a telephone call, send a telegram, or an overnight letter? These are questions that deal more with how we use technologies for our gratification in daily behavior, rather than the details of how we operate them.

A still more encompassing level is to inquire into how more broadly our social or institutional behaviors interact with technologies. Much of the material in the preceding section of the book was on this level, where, for example, we examined the extension of group processes into teleconferencing, or how word processing may change an office culture, or the degree to which microcomputers could be introduced into a school's core curriculum. All inquiries up to this level remain relatively specific; we are dealing with a technology or a group of them and a particular range of behavior. On a much broader level are the generalized consequences of the technologies on our perception of our environment. Put another way, as we are increasingly engaged with an environment that is technologically artifactual, what consequences does it have on the usual parameters of time, space, or speed, what new types of reinforcements might we seek? On this level, the questions are more akin to those that an anthropologist might ask when visiting a remote island: How is people's life space different? Many of the ideas in this chapter were drawn from an earlier conceptual paper by Ronald E. Rice, Herbert S. Dordick, and myself, in which we proposed six areas of broad behavioral impact of the information age. These corresponded to attitudes, perception of time and space, a concept we call connectivity, mobility, increased choice, and socialization. On the one hand, we felt that these were among the more profound environmental and hence, behavioral, impacts of the information age, and we also considered them as key areas for research.

SIX DIMENSIONS OF IMPACT

1. Attitudes

Let us use a traditional psychological concept of attitude, that is, a predisposition to respond, a general expectation of the consequence of one's involvement with a stimulus, a basic evaluation. That we have attitudes about the world around us is nothing new; indeed, it is a fundamental premise of social psychology. What is important in the present discussion is that we tend to acquire somewhat interesting dualistic attitudes toward communication and communication technologies. That is, on the one hand, we have an attitude about the technology itself—e.g., that computers will be intimidating, or that commercial television is trivial, or that data networks are forbiddingly complex, or that electronically synthesized music is bizarre. These we might call media technology (or media) stereotypes, and they may affect not only our thoughts about our medium but also what we expect when we are dealing variously with messages or materials associated with that medium. This reflects Marshall McLuhan's famous phrase, "The medium is the message." If we tend to think that TV is a relatively frivolous medium, we may not expect to view programs of important informational content. Indeed, in our experience with major educational television projects, we found this to be the greatest barrier to the implementation of educational or instructional television in schools. It is difficult to get school trustees and parents to take the medium seriously. That is to say, the medium is so stereotyped that even if a TV message were highly informational or a splendid piece of educational material, its stereotype may prevent users from comprehending these values.

These stereotypes may begin early in our lives, as for example, we (Williams, Coulombe, and Lievrouw, 1983) found when surveying the attitudes of nine- to twelve-year-old children about small computers they had just been introduced to at a Saturday camp. Very clearly, these children not only held but were strongly influenced by such attitudes as favorability of computers, their quality, ease of use, and expense. There are numerous similar examples, although mostly anecdotal, of similar attitudes among employees in business departments that are about to adopt a computing technology.

One important avenue of attitudinal research into technologies has been the degree to which attitudes might be used to predict the adoption of a technology, say, by a group of office workers. One example of this was the research by Ruchinskas and Svenning (1981), who examined a wide range of employee attitudes about video conferencing as contrasted with such available communication alternatives as the telephone, face-to-face, or written messages. As would be assumed in the context of the present discussion, detailed attitudes about these technologies were strongly correlated with the probability of use. It should be said, however, that these attitudes were conditioned somewhat by the types of job demands imposed upon the workers. At best, then, the adoption of an office technology may be a compromise between a person's attitudes and the particular demands or pressures for use in a particular job role.

An important training implication of studies of this type is that getting individuals to use a new technology may require changing old attitudes even before building new ones.

Although traditional theories of attitudes are helpful in conceptualizing the particular role of attitude in technology, it also raises an important caveat. We can never be exactly sure of the relationship of attitudes to behavior. On the one hand, research literature is replete with studies that show consistency of attitude and, particularly in Fishbine's (1975) value expectancy view, that expectations of reward are reliable predictors of behavioral intentions. On the other hand, however, there is still that extra distance between an attitude and actually engaging in behavior. It seems that the study of technologies must grapple with this gap as much as prior studies in the area of attitude research. Still on a theoretical level, we might also ask about the nature of those structures that may link attitudes to technological concepts. Are we researching something superficial and particularistic in the relationship between an attitude and a medium, or are there more basic attitudinal mediators that may act in a more general way in affecting our behavior toward a range of technologies?

2. Time and Space

That communication gives us freedom from limitations of time and space is a proposition that has probably existed since the invention of writing, which allowed us to record messages that could be transported across distances far greater than a human voice could carry or could be preserved in time, whether for minutes or centuries. Modern media, particularly electronic technologies, have further added to these freedoms. Again, Marshall McLuhan stimulated thinking on the issue in the 1960s. One of his key theses was that electronic media would free us from the linear and sequential nature of the printed word. Many contemporary practical examples of consequences of media on time and space can be found without the necessity of referring to McLuhan's abstractions. For example, the choice to rent a video cassette rather than travel to a movie theater is a trade that deals with space. That you can watch the cassette from the middle to the end and watch it six times is a reflection of your increased control over time when you are the direct operator of the medium. This same example could also be extended to random access video disk, in which every scene might be rearranged to suit one's taste, so that a drama is perceived nonlinearly rather than in the usual sequence of the recording (in fact, the clever dramatic writers of the next century will probably create new dramatic forms for nonlinear programming).

The computer-based teleconference is another excellent example that bridges both time and space. While the computer organizes the conference and assembles the remarks at some central point, participants are free to join the conference at any time of the day or night or from any point on the globe (or even off of it!) so long as they have the means for telecommunication with the computer (in most cases, this is a personal computer linked to the telephone work via a modem). This type of asynchronous conference allows individuals to join the group at their convenience, read the agenda, read comments of others, inject their own comments, and then go about their business independent of the conference schedule. Indeed, if one finds major gratifications from interactions with electronic text, it is possible literally to roam the different information services and join conferences or interest groups on a wide variety of topics. In a more localized version of this, enthusiasts operate or participate widely in local electronic bulletin boards. All these represent behaviors that take advantage of an environment freed of time and space.

Space, in particular, has been the subject of increasing attention to the so-called transportation-communication tradeoff. This was involved in the earlier example of bringing the film to your home rather than you to the film, working at home, going to school at home or at work rather than a classroom, or ordering catalog goods from the electronic mall. The prospects of transportation tradeoffs are important to today's urban planner, who seeks to reduce the need of individuals to travel to central locations to pursue their occupations. Well over a decade ago, an inventor, Peter Goldmark (1972) introduced the concept of the "New Rural Society" in which people could dwell in pleasant environments while maintaining their necessary business connections over broadband telecommunications links.

The decentralization of industry is also a topic related to communication-transportation tradeoffs. In a city like Los Angeles, some information intensive industries (banking and insurance) have intentionally divided their work force among a decentralized network of buildings, where each unit is within easy travel distance of a worker population. All of this is much less expensive for the corporation as well as the worker, compared to a centralized, downtown facility. Similarly, countries that are attempting to industrialize in this age but do not want to change their population distribution (e.g., Spain) are seriously investigating the means to decentralize industries, which would in turn be coordinated by broadband networks. All of this saves the cost of transportation, road-building, pollution, and frayed nerves.

But there is a negative side to the time and space concept: Too much freedom may cause us to lose those natural buffers that have traditionally screened out unwanted messages or have discouraged us from wasting our time on extraneous communications. One pressing example of this is the likely negative consequences of media abundance on the lives of our children. Many of today's urban children have a choice of 30 to 50 television channels, vast cassette libraries, on-line information services, not to mention a wide variety of print publications, plus the long-valued telephone. There are far more interesting and gratifying media alternatives that could be used constantly through the waking hours of a child's day if there were not some means for constraining these activities. As it currently seems, part of the communications training of our children is not just how to acquire the right information or entertainment but how to defend oneself against the invasion of all others.

The automated office is another worrisome example. The modern executive is often a victim of information overload, meaning that so much information flows to one's desk that it is impossible to get to the facts or the knowledge that one can use most. Technology has assisted us in generating an abundance of messages and in transporting those messages, but it is just now coming of age and enabling us to sift among them. Thus, it is important in the design of communications systems to incorporate constraints that protect users from extraneous messages, in systems that maintain the desired flow for the organization or the entertainment environment for the home. We seem to be still in an era of unlimited growth of alternatives and only beginning to research and implement methods for mediating choices.

3. Connectivity

One of the most vivid and metaphorical examples of connectivity was essayist Lewis Thomas's reference to the increased population density and human interaction on the earth's surface to the networks on the surface of the living cell. But psychologically and practically speaking, connectivity may most practically refer to one's awareness of potentially being in communication with many different individuals and groups: That you could telephone another telephone user in almost any country in the world in a matter of minutes is a practical example of connectivity. So, too, is that feeling enhanced when one joins an active computer-based teleconference or engages actively in exchange of electronic mail. Examples need not be two-way or interactive, because there is a sense of connectivity (although one-way) engendered in staying in touch with the world of books, the printed press, or the saturated environments of television or radio broadcasting. (Interestingly, research indicates that among the older population, increased media use may be a replacement for a lack of long-time interpersonal contacts—a type of substitute connectivity.)

Connectivity requires an awareness and even an engagement in the present sense of the concept. A person living in a dense urban environment may not sense any greater degree of connectivity than a person in an isolated rural area if there is no active interaction with other individuals or groups. This is where the role of communications technology emerges importantly, because it is the technology that tends to be the growth factor for connectivity—not transportation, not urban change, nor the traditional physical factors.

Beyond refining the concept of connectivity, there are a myriad of questions regarding how people sense connectivity, how they make decisions to increase or reduce it, and whether it may reflect a new type of social relationship that is not simply an extension of interpersonal behavior nor a more involving one of mass media behavior. What is the geography of our social and organizational networks as we mentally conceive of them? What is the interaction between engagement with such networks and our self concept, or at least our concept of the multiple roles within which we might engage? What are the behavioral or social strategies for engaging with a new network?

On a more sociological level is the inquiry into the nature of social structures among individuals or groups not physically proximate. What are the processes by which some of the traditional structural elements such as bridges, cliques, or patterns of various types develop? Are these related to qualities of particular media technologies? How is status within groups defined on a network, or need it be? Will other traditional social differentiations—majorities versus minorities, superiors and subordinates, and high versus low social status—find their way into media-based networks?

Above all, there is the question of how far the individual human can link to the other nodes of a network. There is only so much time in the waking day to be in contact with others, so will there be a limitation of contacts? On the other hand, will capabilities of the media themselves serve to further enlarge one's network, such as was found in the ability to send electronic mail to long lists of addressees rather than just to an individual?

4. Mobility

Just as in the discussion of time and space we emphasized how communications technologies allow us to bring services to our point in space rather than transporting ourselves to the service, so, too, can we see this phenomenon in terms of our increasing freedom for mobility. Advertisements for personal computers, particularly portable or lap computers, are examples of the promotion of this concept. The idea in these examples is that you can continue to do your computer-based work even if you are not at the office but are at home, in a hotel room, on a airplane, or even at the beach. The first-level consequence is that we are no longer as fixed to certain locations, such as home and office, as we once were. Technology allows us the freedom to be almost anywhere we wish, so long as we can maintain contact with the network.

Another example of the technologies promoting mobility is the recent growth of mobile phone services. You can now telephone from your car, the swimming pool, or an airplane. As Herbert Dordick and I have written in our book, *Innovative Management Using Telecommunications,* such mobility changes the way we do business. Lost commuter travel time can be simultaneously reallocated to the placing of routine telephone calls, or the time on a lengthy, multiple-stop trip need not interrupt our capability of checking our electronic mail several times a day. And, of course, such technologies have been well-used by individuals in occupations in which mobility is a key factor, such as in using paging technologies for dispatching service personnel, deliveries, or directing emergency vehicles. There is also the example of the real estate broker whose telephone was literally the lifeline of her business. To be able to answer calls while in a car changed the way she did business.

As Professor Dordick wrote in the essay upon which this chapter is based, "Mobility is embedded in the American psyche." If it is, indeed, the case that there remain few opportunities for increasing our geographic mobility, can we look to electronic communications to provide a substitute? Although this discussion overlaps the earlier one of time and space, it is possible to roam the network in considering the use of databases, electronic mail, and various types of on-line interest groups. In fact, electronic interest groups may reflect a particularly appealing type of community to a population that is enamored of mobility. One needs to make little commitment to an interest group, so it is very easy to move among them, literally and figuratively covering vast numbers of topics and amounts of space.

As we have long known, communication technologies such as the telephone have allowed us to undertake physical displacements from our former social groups and to substitute media contact for face-to-face interaction. It is not difficult to maintain links to family and friends through the telephone network once it becomes possible to move from the neighborhood, across the city, or even to a different part of the country.

Questions raised by the mobility concept include the following: What is the degree of inverse relationship between the availability of communication alternatives and the feeling of freedom to be mobile? Are there subtle, yet important rewards in sensing the freedom of mobility—that is, in the ability to choose more of your mobility pattern apart from earlier constraints of family and social relationships, work, or even recreation? Or put another way, are the psychological problems of mobility (e.g., feelings of noncommitment or rootlessness) likely to be less if a person is engaged in a number of communications-based groups? Finally, are we trading the tradition of geographic mobility in our society for a communications one?

5. Increased Choice

Another distinguishing impact of the new media technologies is that in their many forms they present us with a vast increase in choice. This refers not only to stations on our TV or cable dials and electronic networks but also to the increased use of interactive technologies ranging from the telephone (especially long distance) to electronic mail and conferencing systems. In the business world, for example, there are so many competing media that the ability to satisfy a communication need may require as much attention to the selection of the medium as it does to the structuring of messages.

What is the psychology of media choice? What do we value? As discussed earlier, our pilot study with college-age residential phone users revealed three distinct usage alternatives, each of which might affect choice quite differently; these corresponded to personal contacts (informal "reach out and touch someone" calls), business transactions, and emergencies. Is the use of the telephone so stereotyped that we think of it only in terms of these uses and make our choices accordingly? Put another way, it is probably unlikely that we begin with a major sense of need and then rationally weigh our different media alternatives for gratifying that need. Instead, we may impulsively act upon the basis of media stereotypes, such as to turn on the television for instant entertainment, to search for a definitive book or call an expert when we are in need of facts, or to send a telegram in the hopes that our message will have an impact.

We can also look at choice in the institutional setting, such as when an office decides to adopt a new computer system, to implement word processing, or to install intelligent telephones. Experience reveals that choice in such situations is often based on haphazard factors, such as the desire to be the first on the block with the new technology or to buy a service that everybody else seems to be acquiring, or such as the fact that a major component of one competing system is less expensive than another. We are only now developing deeper insights into the process of institutional choice of information technologies. Given that a functional need has been well-defined, the rational approach is an attempt to uncover all factors that will be relative to the implementation, and this often uncovers such variables as the need for training, the time taken to complete the implementation, required changes in the office environment, and required changes in the work force or management. Some of these factors may ultimately be more costly than the technologies themselves. Nevertheless, implementation will require their inclusion. There is even a broader consideration of choice, such as expressed in a theme of Paul Strassmann's book, *Information Payoff*. Ultimately, according to Strassmann, an information technology is only as valuable to us as it is able to increase our return upon what the market will pay for products or services. In this value-added approach, technology must be assessed not only in terms of how it enhances the development of products or service, but also in terms of our ability to be profitable relative to the market's ability to absorb increased production.

As with any other choice behavior, cost, including economic cost, is important for information technologies. One particularly interesting point, however, is the dispute over whether products should be priced relative to a fair profit margin over their cost of manufacture or whether they should be gauged to the relative worth as envisaged by the buyer. (Certainly, most manufacturers will choose whatever figure is higher.) This has been an issue in the pricing of personal computer software. If a manufacturer can amortize the cost of developing an expensive program across a period of time, the manufacturing costs plus a contribution to that amortization might allow the company to sell a $500 package for, say, a tenth of the price. On the other hand, potential buyers on the market see that software (for example, an integrated package like Lotus 1-2-3) as being of high value to them, then the market price will be established somewhere within the range of that sense of value. Individuals debating the software pricing issue argue that high-priced software is discouraging the growth of personal computer use and hence its own business in the longer run. On the other hand, proponents of the perceived-value position hold that only through high prices will there be sufficient revenue developed to create programs that will attract more buyers to the world of personal computers. Time, of course, will tell. There is also a national concept of choice as regards the acquisition of media technologies by a nation. This has been a particularly salient issue in the case of less developed nations of the world. In many such nations, resources have been invested, for example, in a national television system for which the economic benefits may not be as great as, say, the installation of a telephone system. This is akin to the practice in many third world countries of establishing subsidized airline services simply because of prestige and convenience, even though the budget is disastrous for their economies.

6. Socialization

In the sense that we have been thinking of the new information technologies as modifying our environment, we can also think of those modifications as affecting human learning. Broadly, this is a concept of the impact of information technologies upon socialization. In a world inundated with communication alternatives, what new concepts must a child learn? Are there consequences upon the processes or styles of learning?

Already there is considerable evidence of how children's environments are changing due to media. For example, the AC Nielsen figures for 1984 indicate that children in the two- to five-year age group watch an average of 3.5 hours of television per day and that at 11:30 at night there still are ten million children in this age group who are watching programs. It is also the increasing problem of the lack of control over adult television and film materials as they are now widely available through video-cassette rental and over pay-television channels. It is also a startling contrast to see that youngsters, either at home or at school, have computing machines available to them that once were available only in the form of million-dollar mainframes that could be operated only by white-smocked engineers. There are also the technology-oriented toys which, at least in recent experiences, tend to come and go. During the late 1970s and early 1980s videogames were the fad. (At one point, the annual sales of videogames matched the revenues of the motion picture industry!) By the mid-1980s, pseudoscience toys were the vogue (such as "transformer" toys that could be converted from humanoid figures into cars, trucks, or spaceships). There is also the increasing number of science fantasy shows that fill the Saturday morning hours of children's programming, most of which, authorities have noted, bear little or no relation to the real ways of science. Our point here is not to attempt to survey the details of this changed environment but to offer the generalization that children are growing up in an environment that is dominated by technological concepts, some real and many fantasized.

There are several theoretical positions that enable one to reflect upon the consequences of the changed environment for socialization. As for the consequences of the television environment, probably George Gerbner's *Cultivation* thesis draws the most attention. It asserts that the more that an individual gains concept experience from television as contrasted with real life, the more that the person's expectations of the world will resemble that of television. This means that television socialization engenders such media stereotypes as women with wax buildup on floors, supercops, and six-year-old children who seem extremely clever in dealing with the world (often played by 15-year-old actors). It should be noted, however, that not all researchers agree with the cultivation thesis—for example, sociologist Paul Hirsch (1981), who has analyzed some of the same data as Gerbner and has drawn quite different conclusions.

Another theoretical view centers around the "disappearance" of childhood and is found in the works of such authors as Postman, Winn, Packard, and Elkind. Although these researcher-authors have developed individualized views, all hold that the nature of childhood has changed in our times. In particular, Postman's disappearance thesis is that the media, in particular, have lessened the traditional dividing line between the experiences of children and adults. For example, children can now witness all the adult themes of murder, rape, and backhanded business dealings as portrayed on daily television. Gone is the barrier of print literacy that often separated children from the more lurid themes of adult-oriented print literature. The same themes dominate in the public visual media.

Taken together, the key questions concerning the effects of the environment upon these children as they become adults are the following: What are the consequences for a child's emotional development (a research theme pursued by psychologist Aimee Dorr, at the University of California at Los Angeles)? What are the child's expectations from realistic technologies if most experience has been based upon Saturday morning cartoon fantasies? Are we rearing a generation whose main role models are concoctions of Hollywood writers rather than the outstanding individuals of a child's society?

Topics for Research or Discussion

▬▬▬ The primary thesis of this chapter was that communication and computing technologies are changing the nature of our psychological environment. Examine your own daily life for a one-week period. Identify ways that communication technologies have changed your environment. What do you foresee in terms of further changes, if any?

▬▬▬ As we generalize in several places in the present book, most new media are extensions of old ones. Construct an outline or figure that shows the relation of examples of new media to traditional media. From an overall view of your analysis, answer such questions as: What are some of the contrasts between new and old media? What do you think the psychological consequences of new media are upon the user? Do you think that still newer forms of media will be developed, and if so, what might they be?

▬▬▬ Marshall McLuhan gained international recognition in the 1950s and 1960s as a teacher and commentator on modern electronic media. Review some of McLuhan's contributions. One of his famous statements was that "the medium is the message." What do you think that he meant by this? What did he mean by hot versus cold media? (TIP: His two important books were *The Gutenberg Galaxy,* Toronto: University of Toronto Press, 1951 and *Understanding Media,* New York: McGraw-Hill, 1964.)

▬▬▬▬ With the abundance of audiovisual media now available to us, some critics have decried the loss of public interest in print communication. Do you feel that modern society is less oriented to print media than before? What do you think is the future of print? (Be sure to include within your answer your interpretation of print as it appears in text form in broadcast or computer media.)

▬▬▬▬ As new media technologies have allowed media choice and contact to become more personalized, this process has been characterized as demassification. Prepare a brief report on your interpretation of demassification and include several contemporary examples, especially ones that you may have personally experienced. Finally, comment upon what you consider to be future trends in demassification.

References and Further Readings

Bell, D. *The Coming of Post-Industrial Society*. New York: Basic Books, 1976.

Bennis, W., and P. Slater. *The Temporary Society*. New York: Harper & Rowe, 1968.

Bernard, H., and P. Killworth. "Informant Accuracy in Social Network Data II." *Human Communication Research* 4 (1977):3–18.

Bowers, B. *Communications for a Mobile Society: An Assessment of New Technology*. Beverly Hills, Calif.: Sage, 1978.

Comstock, G., S. Chaffee, N. Katzman, M. McCombs, and D. Roberts. *Television and Human Behavior*. New York: Columbia University Press, 1978.

Cowell, R., and R. Wigand. "Communication Interaction Patterns among Members of Two International Agricultural Research Institutes." Paper presented at the annual conference of the International Communication Association, Acapulco, Mexico, 1980.

Delaney, J. "The Efficiency of Sparse Personal Contact Networks for Donative Transfer of Job Vacancy Information." University of Minnesota, Department of Sociology, Working Paper 80-03, 1980.

deMause, L. "The Evolution of Childhood." In *The History of Childhood*, edited by L. deMause. New York: The Psychohistory Press, 1974.

Dordick H. S., H. G. Bradley, and B. Nanus. *The Emerging Network Marketplace*. Norwood, N.J.: Ablex, 1981.

Dordick, H. S., P. Lum, and A. Phillips. "Social Uses for the Telephone." *InterMedia* 11(3) (1983):31–34.

Dordick, H. S., and F. Williams. *Innovative Management Using Telecommunications*. New York: John Wiley, 1986.

Elkind, D. *The Hurried Child*. Reading, Mass.: Addison-Wesley, 1983.

Gailbraith, J. *Organization Design*. Menlo Park, Calif.: Addison-Wesley, 1977.

Gerbner, G., L. Gross, N. Signorielli, M. Morgan, and M. Jackson-Beeck. "The Demonstration of Power: Violence Profile Number 10." *Journal of Communication* 26(1979): 173–199.

Gershuny, J. *After Industrial Society: The Emerging Self-Service Economy*. Atlantic Highlands, N.J.: Humanities Press, 1978.

Ghez, G., and G. S. Becker. *The Allocation of Time and Goods Over the Life Cycle*. New York: Columbia University Press, 1975.

Glossbrenner, A. *The Complete Handbook of Personal Computer Communications*. New York: St. Martin's Press, 1983.

Goldmark, P. C. "Tomorrow We Will Communicate to Our Jobs." *The Futurist* 6, 2 (1972): 35–42.

Gottman, J. "Megalopolis and Antipolis: The Telephone and the Structure of the City." In *The Social Impact of the Telephone*, edited by I. Pool. Cambridge, Mass.: MIT Press, 1977.

Hiltz, S. R., and M. Turoff. *The Network Nation*. Menlo Park, Calif.: Addison-Wesley, 1978.

Hirsh, P. M. "The 'Scary World' of the Nonviewer and Other Anomalies: The Reanalysis of Gerbner et al.'s Findings on Cultivation Analyses, Part I." *Communication Research* 7(1980): 403–456.

Korte, C., and S. Milgram. "Acquaintance Networks between Racial Groups: Applications to the Small World Problem." *Journal of Personality and Social Psychology* 15 (1970): 101–108.

Long, L., and Boertlin. "The Geographical Mobility of Americans." *Current Population Reports, Special Studies Series,* P-23, No. 64. U.S. Bureau of the Census. Washington, D.C.: U.S. Government Printing Office, 1976.

MacCannell, D. *The Tourist: A New Theory of the Leisure Class*. New York: Schocken, 1976.

McLuhan, M., and Q. Fiore. *The Medium Is the Message*. New York: Bantam, 1967.

Niles, J. et al. *The Telecommunications–Transportation Trade Off*. New York: John Wiley, 1976.

Packard, V. *Our Endangered Children: Growing Up in a Changing World*. New York: Little, Brown, 1983.

Parsons, T. "The Social Structure of the Family." In *The Family: Its Functions and Destiny,* edited by R. Ansker. New York: Harper & Row, 1949.

Pattison, R. *On Literacy*. Oxford: Oxford University Press, 1982.

Phillips, A. *Attitude Correlates of Selected Media Technologies: A Pilot Study*. Los Angeles: Annenberg School of Communications, 1982.

Phillips, A. "Computer Conferences: Success or Failure?" Vol. 7, *Communication Yearbook,* edited by R. Bostrom. Beverly Hills, Calif.: Sage, 1983, 837–856.

Phillips, A., P. Lum, and D. Lawrence. "A Conceptual Framework for the Cross-cultural Study of Telephone Use." Paper presented at the Conference on Communications and Culture, Temple University, Philadelphia, March 1983.

Pool, I., ed. *The Social Impact of the Telephone.* Cambridge, Mass.: MIT Press, 1977.

Postman, N. *The Disappearance of Childhood.* New York: Delacorte Press, 1982.

Rapoport, A. "A Probabilistic Approach to Networks." *Social Networks* 2(1) (1979): 1–18.

Reid. A. L. "Comparing Telephone with Face-to-Face Contact." In *The Social Impact of the Telephone,* edited by I. Pool. Cambridge, Mass.: MIT Press, 1977.

Rice, R. E. "Impacts of Organizational and Interpersonal Computer-Mediated Communication." Vol. 16, *Annual Review of Information Science and Technology,* edited by I. Pool. White Plains, N.Y.: Knowledge Industry Productions, 1980, 221–249.

Rice, R. E. "Communication Networking in Computer-Conferencing Systems: A Longitudinal Study of Group Roles and System Structure." Vol. 6, *Communication Yearbook,* edited by M. Burgoon. Beverly Hills, Calif.: Sage, 1982, 925–944.

Rice, R. E., and Associates. *The New Media: Communication, Research and Technology.* Beverly Hills, Calif.: Sage, 1984.

Rice, R. E., and W. Paisley. "The Green Thumb Videotext Project: Evaluation and Policy Implications." *Telecommunications Policy* 6(3) (1982): 223–236.

Rice, R. E., and W. Richards, Jr. "An Overview of Communication Network Analysis Methods." In Vol. 6, *Progress in Communication Sciences,* edited by B. Dervin and M. Voigt. Norwood, N.J.: Ablex, 1984.

Robbins, L. *An Essay on the Nature and Significance of Economic Science.* London: Macmillan, 1945.

Robinson, J., and P. Converse. "The Impact of Television on Mass Media Uses: A Cross National Comparison." In *The Use of Time,* edited by A. Syalai. The Hague: Mouton, 1972.

Rogers, E. M. *Diffusion of Innovations.* New York: Free Press, 1983.

Rogers, E. M., and R. Agarwala-Rogers. *Communication in Organizations.* New York: Free Press, 1976.

Rogers, E. M., H. M. Daley, and T. D. Wu. *The Diffusion of Home Computers.* Stanford, Calif.: Institute for Communication Research, Stanford University, 1982.

Rogers, E. M., and L. Kincaid. *Communication Networks*. New York: Free Press, 1981.

Ruchinskas, J. "Communicating in Organizations: The Influence of Context, Job, Task, and Channel." Ph.D. dissertation, University of Southern California, 1982.

Ruchinskas, J., and L. Svenning. "Formative Evaluation for Designing and Implementing Organizational Communication Technologies: The Case of Video-conferencing." Paper presented at the annual conference of the International Communication Association, Minneapolis, May 1981.

Rytina, S., and D. Morgan. "The Arithmetic of Social Relations: The Interplay of Category and Network." *American Journal of Sociology* 88(1) (1982): 88–113.

Sharp, C. *The Economics of Time*. New York: John Wiley, 1981.

Short, J., E. Williams, and B. Christie. *The Social Psychology of Telecommunications*. New York: John Wiley, 1976.

Simon, H. "Applying Information Technology to Organizational Design." *Public Administration Review* 33(3) (1973): 268–278.

Singer, B. "Crazy Systems and Kafka Circuits." *Social Policy* 11(2) (1980a): 46–54.

Singer, B. "Incommunicado Social Machines." *Social Policy* 8(3) (1978): 88–93.

Singer, B. *Social Functions of the Telephone*. Palo Alto, Calif.: R & E Associates, 1980b.

Stein, M. R. *The Eclipse of Community: An Interpretation of American Studies*. New York: Harper & Row, 1966.

Svenning, L. "Individual Response to an Organizationally Adopted Telecommunications Innovation: The Difference among Attitudes, Intentions, and Projections." Paper presented at the annual conference of the International Communication Association, Dallas, May 1983.

Svenning, L. "Predispositions toward a Telecommunication Innovation: The Influence of Individual, Contextual, and Innovation Factors on Attitudes, Intentions, and Projections toward Video-conferencing." Ph.D. dissertation, University of Southern California, 1982.

Svenning, L., and J. Ruchinskas. "Organizational Teleconferencing." In *The New Media: Uses and Impacts*, edited by R. Rice. Beverly Hills, Calif.: Sage, 1984.

Von Neumann, I., and O. Morgenstern. *Theory of Games and Economic Behavior*. New York: John Wiley, 1964.

Weick, K. *The Social Psychology of Organizing*. Menlo Park, Calif.: Addison-Wesley, 1969.

Whyte, W. F. *Street Corner Society*. Chicago: University of Chicago Press, 1955.

Wiener, N. *Cybernetics: Or Control and Communication in the Animal and the Machine*. New York: John Wiley, 1948.

Williams, E. "Experimental Comparison of Face-to-face and Mediated Communications: A Review." *Psychological Bulletin* (1977): 963–976.

Williams, F., J. Coulombe, and L. Lievrouw. "Children's Attitudes toward Small Computers: A Preliminary Study." *Educational Communication and Technology* 31(1) (1983).

Williams, F., and H. Dordick. *The Executive's Guide to Information Technology*. New York: John Wiley, 1983.

Williams, F., A. Phillips, and P. Lum. *Some Extensions of Uses and Gratifications Research*. Los Angeles: Annenberg School of Communications, 1982.

Williams, F., and R. E. Rice. "Communication Research and the New Media Technologies." Vol. 6, *Communication Yearbook,* edited by R. Bostrom. Beverly Hills, Calif.: Sage, 1983, 200–224.

Williams, F., and V. Williams. *Microcomputers in Elementary Education*. Belmont, Calif.: Wadsworth, 1984.

Winn, M. *Children Without Childhood*. New York: Pantheon, 1983.

Glossary

access channel Cable television channel dedicated to public use, often with provisions for the general public to originate its own programs but sometimes only for governmental or educational purposes.

alphanumeric Digital and alphabet characters used in computer text or code.

AM Amplitude modulation (*see* modulation).

amplifier Process that takes a signal and increases its power by drawing power from a source other than the signal itself.

analog Representation that bears some physical relationship to the original quality: usually electrical voltage, frequency, resistance, or mechanical translation or rotation.

antenna Device to collect and/or radiate radio energy.

artificial intelligence Computer programs that perform functions, often by imitation, usually associated with human reasoning and learning.

ASCII (Pronounced "ask-ee.") American Standard Code for Information Interchange. Binary transmission code used by most teletypewriters and display terminals.

band Range of radiofrequencies within prescribed limits of the radiofrequency spectrum.

bandwidth Width of an electronic transmission path or circuit in terms of the range of frequencies it can pass; a measure of the volume of communications traffic that the channel can carry. A voice channel typically has a bandwidth of 4000 cycles per second; a TV channel requires about 6.5 megahertz (MHz).

baseband Information or message signal whose content extends from a basic frequency that is near some finite value. For voice, baseband extends from 300 hertz (Hz) to 34,000 Hz; video baseband is from 50 Hz to 4.2 MHz (NTSC standard).

Some of these definitions are from earlier works by the author, namely: The New Communications *(Belmont, Calif.: Wadsworth, 1983) and* The Executive's Guide to Information Technology *(with H.S. Dordick; New York: John Wiley & Sons, 1983).*

baud Bits per second (bps) in a binary (two-state) telecommunications transmission.

binary Numbering system having only two digits, typically 0 and 1.

bit Binary digit. Smallest piece of information, with values or states of 0 or 1, or yes or no. In electrical communication system, a bit can be represented by the presence or absence of a pulse.

booster Amplifier in a communications system that increases the power of a signal for retransmission to a further point in the system.

bps Bits per second (*see* bit).

broadband carriers High-capacity transmission systems used to carry large blocks of, for instance, telephone channels or one or more video channels. Such broadband systems may be provided by coaxial cables and repeated amplifiers or microwave radio systems.

broadband communication Communications system with a bandwidth greater than voiceband. Cable is a broadband communication system with a bandwidth usually from 5 MHz to 450 MHz.

buffer Machine or other device inserted between other machines or devices to match systems or speeds, prevent unwanted interaction, or delay the rate of information flow.

byte Group of bits processed or operating together. Bytes are often an 8-bit group, but 16-bit and 32-bit bytes are not uncommon.

cable television Use of a broadband cable (coaxial cable or optical fiber) to deliver video signals directly to television sets, in contrast to over-the-air transmission. Current systems may have the capability of receiving data inputs from the viewer and of transmitting video signals in two directions, permitting pay services and video conferencing from selected locations.

CAD Computer-aided design. Techniques that use computers to help design machinery and electronic components.

CAI Computer-assisted instruction.

CAM Computer-aided manufacturing.

capacitance Ability of a system of layers of conductors and nonconductors to store electrical energy.

carrier Signal with given frequency, amplitude, and phase characteristics that is modulated in order to transmit messages.

cassette Enclosed reel for spooling electromagnetic tape; developed for ease of use over open reel systems.

cathode ray tube (CRT) Display unit or screen for a computer.

CATV Community antenna television; term used to refer to the forerunner of cable television systems.

cellular radio (telephone) Radio or telephone system that operates within a grid of low-powered radio sender-receivers. As a user travels to different locations on the grid, different receiver-transmitters automatically support the message traffic. This is the basis for modern cellular telephone systems.

CCITT Consultative Committee for International Telephone and Telegraphs, an arm of the International Telecommunications Union (ITU) that establishes voluntary standards for telephone and telegraph interconnection.

central office Local switching center for a telephone system, sometimes referred to as a wire center.

channel Segment of bandwidth that may be used to establish a communications link. A television channel has a bandwidth of 6 MHz, a voice channel about 4000 Hz.

character generator Device that creates alphanumeric characters for display on a cathode ray tube (or television) screen.

chip Single device made up of interconnected transistors, diodes, and other components.

circuit switching Process by which a physical interconnection is made between two circuits or channels.

coaxial cable Metal cable consisting of a conductor surrounded by another conductor in the form of a tube that can carry broadband signals by guiding high-frequency electromagnetic radiation.

common carrier Organization licensed by the Federal Communications Commission (FCC) and/or by various state public utility commissions to supply communications services to all users at established and stated prices.

communications settings Whatever is necessary for two computers to use the same communication system; often called "protocol."

compiler Automatic computer coding system that generates and assembles a program from instructions prepared by equipment manufacturers or software companies.

computer word String of characters or binary numbers considered as one unit and stored at a single computer address or location.

COMSAT Communications Satellite Corporation. A private corporation authorized by the Communications Satellite Act of 1962 to represent the United States in international satellite communication and to operate domestic and international satellites.

converter Device that makes the necessary frequency transformations so that signals on cable TV channels will be compatible with home television sets.

CPU Central processing unit of a computer.

CRT *See* cathode ray tube.

database Information or files stored in a computer for subsequent retrieval and use. Many of the services obtained from information utilities involve accessing large databases.

DBS *See* direct broadcast satellite.

decoder Device used to unscramble a communication signal, such as a pay-TV channel, so that it can be viewed normally.

dedicated lines Telephone lines leased for a specific term between specific points on a network, usually to provide certain special services not otherwise available on the public watched network.

demodulate Process of transforming a particularly modulated wave so that it will appear in its original form.

descrambler *See* decoder.

digital Function that operates in discrete steps, as contrasted with a continuous or analog function. Digital computers manipulate numbers encoded into binary (on-off) forms, while analog computers sum continuously varying forms.

digital communications Transmission of information using discontinuous, discrete electrical or electromagnetic signals that change in frequency, polarity, or amplitude. Analog messages may be encoded for transmission on digital communication systems (*see* pulse code modulations).

direct broadcast satellite (DBS) Satellite system designed with sufficient power so that inexpensive earth stations can be used for direct residential or community reception, thus reducing need for local networks.

dish Satellite earth station.

disk (or disc) Magnetized surface capable of storing binary information.

disk operating system Operating program that instructs the computer how to store and retrieve information from a disk.

DOS *See* disk operating system.

dot matrix Computer-driven printer that prints symbols with a matrix of small pins (as distinguished from a letter-quality printer).

down converter Equipment for transforming high-frequency signals to low-frequency ones, as in satellite downlinks.

downlink Antenna designed to receive signals from a communications satellite (*see* uplink).

download To receive information from another computer and store it into a computer memory or disk files.

downstream Signals sent from a cable television broadcasting center to homes.

drop Link that connects a subscriber's television set in the home to the outside cable TV system.

dumb terminal *See* terminal.

duplex Two-way communication (information can flow simultaneously both ways in a communication link). Often called "full duplex" as contrasted with one-way communication or "half duplex."

earth station Communication station on the surface of the earth used to communicate with a satellite. Also, television receive-only earth station (TVRO).

electromagnetic spectrum Full range of radiant energy frequencies, which includes light and radio frequencies.

electronic mail Delivery of correspondence including graphics by electronic means, usually the interconnection of computers, word processors, or facsimile equipment.

encryption Change from a plain text to an encoded form that requires sophisticated techniques for decoding. Digital information can be encrypted directly with computer software.

ESS Electronic Switching System. Bell System designation for their stored program control switching machines.

facsimile System for the transmission of images. Black-and-white reproduction of a document or a picture transmitted over a telephone or other transmission system.

fax *See* facsimile.

FCC Federal Communications Commission. Board of five members (commissioners) appointed by the President and confirmed by the Senate under the provision of the Communications Act of 1934. Has the power to regulate interstate communications.

fiber optics Glass strands that allow transmission of modulated light waves for communication.

file (computer) Information stored in a computer. Large files that contain information for retrieval are typically called databases.

final mile Link in a satellite telecommunications system from the earth station to wherever programs are originated or viewed. May involve coaxial cable, microwave relay, or other types of terrestrial communication systems.

firmware Instructions for operation of computer or communications equipment that are embedded in the electrical design of a microchip.

floppy disk Plastic disk used to store computer information or programs (*see* disk).

FM Frequency modulation (*see* modulation).

footprint Area of the earth that receives signals from an individual satellite.

frequency Number of recurrences of a phenomenon during a specified period of time. Electrical frequency is expressed in hertz, equivalent to cycles per second.

frequency spectrum Range of frequencies of electromagnetic waves in radio terms; range of frequencies useful for radio communication, from about 10 kilohertz to 3000 gigahertz.

full duplex *See* duplex.

gateway Ability of one information service to transfer you to another one, as from Dow Jones News/Retrieval to MCI Mail.

geosynchronous satellite Satellite with a circular orbit 22,400 miles in space that lies in the plane of the earth's equator. Turns about the polar axis of the earth in the same direction and with the same period as the earth rotates. Thus the satellite is stationary when viewed from the earth.

gigahertz (GHz) Billion cycles per second.

half duplex Message flow that is only one-way at a time (*see* duplex).

handshaking Electronic exchange of signals as one computer links with another.

hard copy *See* printout.

hardware Electrical and mechanical equipment used in telecommunications and computer systems (*see* software; *also* firmware).

head end Electronic control center of the cable television system where weaving signals are amplified, filtered, or converted as necessary. Usually located at or near the antenna site.

hertz (Hz) Frequency of an electric or electromagnetic wave in cycles per second, named after Heinrich Hertz, who detected such waves in 1883.

high-density TV Television picture scanning systems that are greater than the present European or American (525-line) standards, usually over 1000 lines.

holograph Method of image reproduction on a flat surface that gives the appearance of three dimensions.

host Main computer or computer system that is supporting a group of users.

hub Component in a cable TV network that links subscribers in a particular area to a central facility that receives signals from a variety of sources (*see* head end).

IEEE Institute of Electrical and Electronic Engineers. A professional society.

information utility Services that offer a wide variety of information, communications, and computing services to subscribers; examples are The Source, CompuServe, and Dow Jones News/Retrieval.

I/O Input/output. Equipment or processes that transmit data into or out of a computer's central processing unit.

institutional loop Separate cable for a CATV system designed to serve public institutions or businesses, usually with two-way video and data services.

interface Devices that operate at a common boundary of adjacent components or systems and that enable these components or systems to interchange information.

IPS Inches per second.

ITFS Instructional Television Fixed Service. Broadcasting spectrum set aside for educational use.

K 1024 bytes of information, or roughly the same number of symbols or digits.

kilohertz (kHz) Thousand cycles per second.

large-scale integration (LSI) Single integrated circuits that contain more than 100 logic circuits on one microchip (*for comparison see* very large scale integration).

laser Light amplification by simulated emission of radiation. Intense beam of light that can be modulated for communications.

LED Light emitting diode. Type of semiconductor that can be used to create visual images.

local area network Special linkage of computers or other communications devices into their own network for use by an individual or organization. Local area networks are part of the modern trend of office communication systems.

local loop Wire pair that extends from a telephone central office to a telephone instrument. Coaxial cable in a broadband or CATV system which passes by each building or residence on a street and connects with the trunk cable at a neighborhood node. Often called the subscriber loop.

low-power television station Station operating at less than normal power, using special broadcast frequencies so that it will not interfere with existing assigned frequencies. Developed to increase opportunities for television station operation.

LPTV *See* low-power television station.

LSI *See* large-scale integration.

mainframe Base or main part of a large computer, as contrasted with mini or micro computers. Usually refers to the actual processing unit.

mass storage Computer disk or tape device that can hold very large amounts of information cheaply with automated access on demand. A disk unit on a personal computer provides for mass storage.

MATV Master antenna television. System in which one antenna is used for a neighborhood or an apartment and then links to individual dwellings via wired network.

MDS *See* multipoint distribution service.

megahertz (MHz) Million cycles per second.

memory One of the basic components of a central processing unit (CPU). It stores information for future use.

microchip Electronic circuit with multiple solid-state devices engraved through photolithographic or microbeam processes on one substrate.

microcomputer Set of microchips that can perform all the functions of a digital stored-program computer (*see* microprocessor).

microsecond One millionth of a second.

microprocessor Microchip that performs the logic functions of a digital computer.

microwave Short wave lengths from 1 GHz to 30 GHz used for radio, television, and satellite systems.

millisecond One thousandth of a second.

minicomputer In general, a stationary computer that has more computer power than a microcomputer but less than a large mainframe computer.

modem Modulator-demodulator. Equipment that you use to link a computer to a telephone line.

modulation Process of modifying the characteristics of a propagating signal, such as a carrier, so that it represents the instantaneous changes of another signal. The carrier wave can change its amplitude (AM), its frequency (FM), its phase, its duration (pulse code modulation), or combinations of these.

monitor (video) Video screen on a computer, but has more technical meanings as well.

mouse Simple device for moving a marker on a computer screen in order to give the computer commands.

MS-DOS Operating system standardized by IBM personal computers.

MSO Multiple system operator. Company that owns multiple cable TV services.

multiplexing Process of combining two or more signals from separate sources into a single signal for sending on a transmission system, from which the original signals may be recovered.

multipoint distribution service (MDS) Special one-way microwave radio transmission system; mainly known as a means for providing pay-television broadcasting.

NABTS North American Broadcast Teletext Standard (proposed by AT&T).

nanosecond One billionth of a second.

narrowband communication Communication system capable of carrying only voice or relatively slow-speed computer signals.

NCTA National Cable Television Association.

network Circuits over which computers or other devices may be connected with one another, such as telephone network or computer networking.

node Point at which terminals and a computer or computers are connected to the transmission network.

NTSC National Television System Committee (the U.S. system of television standards).

off-line Equipment not connected to a telecommunications system or another operating computer system.

OFS Operations Fixed Service. Microwave service used particularly for business communication.

on-line Connected to a telecommunications or computing system.

operating system Instructions for a computer that permit it to run various programs and handle scheduling, control of printers, terminals, memory devices, and so forth.

optical fiber Thin, flexible glass fiber (the size of a human hair) that can transmit light waves capable of carrying large amounts of information.

orbit Path of a satellite around the earth.

PABX Private Automatic Branch Exchange. Private telephone switching system that provides access to and from the public telephone system. Also, PBX (private branch exchange).

packet switching Technique for switching digital signals with computers, in which the signal stream is broken into "packets" and reassembled in the correct sequence at the destination.

peripherals Units that operate in conjunction with a computer but are not part of it, such as, printers, modems, or disk drives.

personal computer Microcomputer that tends to be used more for business purposes than for games or recreation, but the distinction is not all that exact.

PBX Private Branch Exchange. Telephone switching system.

PLP Presentation Level Protocol. Proposed standard for broadcast for wired text systems (AT&T).

port Place of communication signal entrance to or exit from a computer. (Also sometimes used as a verb in describing the transfer of a computer program from one system to another.)

printout Materials listed by your printer; sometimes called hard copy.

program Set of instructions arranged in proper sequence for directing a computer to perform desired operation.

protocol Description of the requirements for enabling one computer to communicate with another.

public switched telephone network The more formal name given to the commercial telephone business in the United States; includes all the operating companies.

pulse code modulations (PCM) Technique by which a signal is sampled periodically and each sample quantized and transmitted as a digital binary code.

random access Capability of retrieving a select portion of information from among a large body of information without having to proceed through it sequentially.

RAM Random access memory. Provides access to any storage or memory location point directly by means of vertical and horizontal coordinates. Is erasable and reusable.

robotics Use of electronic control techniques, as programmed on microprocessors and microcomputers, to operate mechanical sensing and guidance mechanisms in manufacturing and assembly processes.

ROM Read only memory. Permanently stored memory that is read out and not altered in the operation.

satellite Usually refers to a communication satellite, essentially a radio receiver and transmitting device orbiting the earth and capable of broadcasting signals over a wide area.

scramble Encode a signal so it cannot be decoded by unauthorized users; often applied to television services.

slow-scan television Technique of placing video signals on a narrowband circuit, such as telephone lines, that results in a picture changing every few seconds.

SMATV Satellite master antenna television. Use of satellite dishes on multiple dwellings to capture signals that are then transmitted over wired to residents.

SMPTE Society of Motion Picture and Television Engineers.

software Written instructions that direct a computer program. Any written material or script for use on a communications system or the program produced from the script.

solid state Devices that enable transformations or manipulations of small electrical currents due to the conductive properties of the materials.

STV Subscription television.

subcarrier Signal accompanying a carrier signal that is used to transmit special information and which will require a special receiver (text captions for the deaf are sent in this manner along with a television signal).

systems program Supports the basic operating system of the computer (for example, in allocating memory storage) and operating peripherals, in contrast with an applications program which accomplishes a specific task such as word processing.

tariff Published rate for a service, equipment, or facility established by the communication common carrier.

telecommuting Use of computers and telecommunications to enable people to work at home. More broadly, the substitution of telecommunications for transportation.

teleconference Visual, sound, or text interconnection that allows individuals in two or more locations to see and talk to one another in a long-distance conference arrangement.

telemarketing Method of marketing that emphasizes the creative use of the telephone and other telecommunication systems.

telemetry Transmission of information or messages that are originated by measurement devices.

teletext Generic name for a set of systems that transmit alphanumeric and simple graphical information over the broadcast (or cable) signal, using spare line capacity in the signal for display on a suitably modified TV receiver.

telex Dial-up telegraph service.

terminal Point at which communication can either leave or enter a communication network. (A "dumb" terminal is distinguished from one with built-in computing capabilities.)

terminal emulator Use of a personal computer to act as a "dumb terminal"; usually requires a special program.

timesharing Use of one computer by two or more users. Large computers used by the information utilities can accommodate many users simultaneously who are said to be "timesharing" on the system.

translator Usually the retransmission of a television signal to improve reception in a given area.

transmitter Equipment that creates broadcast carrier waves and imposes information by modulation upon it.

transponder Electronic circuits of a satellite that receive a signal from the transmitting earth station, amplify it, and transmit it to earth at a different frequency.

trunk Main cable that runs from the head end to a local node and then connects to the drop to a home (in a cable television system); main circuit connected to local central offices with regional or intercity switches (in telephone systems).

TVRO In satellite communications, a television receive-only earth station.

twisted pair The two wires that connect local telephone circuits to the telephone central office.

UHF Ultra high frequency.

UNIX Computer operating system developed by AT&T; now popular for small computers that support multiple terminals.

uplink Communications link from the transmitting earth station to the satellite.

upload Transfer information out of the memory or disk file of your computer to another computer.

upstream Information flow from the home to the cable television broadcaster.

user friendly Popular term that means that a computer, a program, or a system is easy to learn or easy to use.

VCR Video cassette recorder.

very large scale integration (VLSI) Single integrated circuits that contain more than 100,000 logic gates on one microchip (*for comparison see* large-scale integration).

VHF Very high frequency.

video monitor Visual display screen of a computer.

videotext Generic name for a computer system that transmits alphanumeric and simple graphics information over the ordinary telephone line for display on a video monitor.

VLSI *See* very large scale integration.

WATS Wide Area Telephone Service. Service offered by telephone companies in the U.S. that permits customers to make dial calls to telephones in a specific area for a flat monthly charge, or to receive calls "collect" at a flat monthly charge.

Index

needs assessments, 111–12
newspaper(s)
    advertising revenues for, 158
    business revenues for, 156
    computer use by, 136, 139–40
    facsimile transmission by, 141–42
    future technologies for, 142–44
    satellite use by, 140–41
    statistics on world, 137
New York Times Co., 142
Nickelodeon, 144
Nielsen Company, 155
Nilles, Jack, 161
Nippon Telephone and Telegraph (NTT),
    148
Northern Ontario Telemedicine System,
    171–75

office technologies, 110
    attitudes toward, 240–41
    organization adoption of new, 117–22
    in organizational communication, 108–
        10
    patent office test program of, 125–28
    promises of, 110–12
    research topics on, 128–29
    studying implementation of, 112–14
optical transmission, 29
organization(s)
    adoption of new office technologies
        by, 117–22
    as communicating entity, 108, 109
    as communications environment, 104–
        7
Otis, Harrison Gray, 166

Packard, V., 248
Paisley, William, 161, 167
Palmgreen, P., 222, 225
Pappert, Seymour, 197
Pascal, 65, 184
Pascal, Blaise, 24, 65
PEACESAT, 212
Pease, Pamela, 95
Pelton, Joseph, 4
personal computers. *See* computers, per-
    sonal
Phillips, Amy F., 79, 93, 95, 226, 231
PL/1, 65
Poratt, Marc, 4, 115

ports, 61
postindustrialism, 203, 205–6
    See also Bell, Daniel
Postman, N., 248
PresTel, 143, 148
productivity
    increasing office, 117–18
    research topic on, 128
Program Evaluation and Analysis Com-
    puter (PEAC), 188
programming, languages for, 64–66
programs, systems and applications, 66–
    67
publishing
    new entries in, 143–44
    revenues in U.S. magazine and book,
        156
    statistics on world book, 137

Qube, 25, 30, 48, 229

Radin, George, 65
radio
    advertising revenues for, 158
    spectrum allocation for, 50
    statistics on transmitters and receivers
        for, 138–39
    transistor, and Third World, 208–9
    world availability of, 203
    world statistics on, 204
RAM (random access memory), 61–62
Ramo, Simon, 4
Rayburn, J. B., 225
RCA Company, 35
receivers, 11–13
reinvention, 113–14
    research topic on, 129
    and word processing, 122–24
Rice, Ronald E., 112, 114, 122, 231,
    239
robots, research topic on, 71
Rogers, Everett M., 202, 207, 208, 215
ROM (read only memory), 61–62
Rosenblatt, Bruce, 65
Ruchinskas, J., 94, 99, 240

Satellite Instructional Television Experi-
    ment (SITE), 209
satellites, 25
    communication, 26–29

and developing countries, 211, 212
newspaper use of, 140–42
spectrum allocation for, 50
and superstations, 144–45
and Third World television, 209
Saturn project, 115
schools. *See* education
Schramm, Wilbur, 202, 207
Schwartz, Jules, 65
Secom, 148
Second Self, The, 197
secretaries, 119
in patent office test case, 125–28
Sesame Street, 181, 187
Shaver, T. L., 227
Short, John, 78
Showtime, 144
Silicon Valley Fever, 215–16
SMATV (satellite master antenna TV), 145
social presence, 76–78
Social Psychology of Telecommunications, 235
Social Research and the Telephone, 87
socialization
media implications for, 190–92
technology's impact on, 247–48
software, 67
study of educational, 192–96
Sony Corporation, 35, 148
Source, The, 33, 78, 92, 229
research topic on, 152
sources, 11–13
space, 7, 241–43
spectrum allocation, 50
Steinfield, Charles, 78, 84, 232
Strassmann, Paul, 114, 116, 125, 246
Subscription TV (STV), 144
supercomputers, 37–39
superstations, 144–45
Svenning, L., 94, 99, 240
switching
circuit, 54
packet, 55
simple networks and, 54

Tandy Corporation, 37
teachers, 182, 194–95
technology(ies)
behavioral impacts of, 239

challenges of, 5–6
and change, 3–6
dimensions of impacts of, 240–48
information, 27, 28
newspaper, 136, 139–42, 142–44
research topics on, 7–8, 40–41, 87, 248–49
theory of, as investment for development, 207
topics for research on new, 232–34
for various types of communication, 19
See also communication technologies; media technologies; office technologies
telecommunications, 11, 45, 46
business revenues from, 157
common systems of, 48
examples of technologies of, 26–36
from ecology of games perspective, 162–67
forming environment, 55
generalized model of, 47
interactive capabilities of, 52–53
media of, 49–52
planning infrastructures for, 213–15
point-to-point, 78–79
processes of, 46–47
research topics on, 56
switching in, 54–55
telecommuting, 5, 161
teleconferencing, 53, 91, 108
generalizations regarding, 91–94
predicting use of, 93–94
research on computer-mediated, 95–98
research topics on, 99–100
and time and space, 241–42
trends in, 94
types of, 92
and uses and gratifications theory, 231
telemarketing, 5
telemedicine, successful project in, 170–75
telephone, 24, 25
cellular mobile, 32
and developing countries, 212
gratifications from use of, 80–83
national wealth and presence of, 203
research topics on, 56, 86–87, 99, 129, 176

To the reader:

I hope that you have had a positive communication experience reading *Technology and Communication Behavior*. The importance of feedback for increasing the effectiveness and efficiency of communications has been discussed at many points in this book. So it should come as no surprise to you that I am interested in receiving feedback from you. I look forward to reading your comments and suggestions and hope to incorporate them in future editions of *Technology and Communication Behavior*. Thank you.

School _____ Instructor's Name _____

Course Title and Number _____

Intended Major _____

1. What features did you like *most* about *Technology and Communication Behavior*? _____

_____

2. What features did you like *least* about *Technology and Communication Behavior*? _____

_____

3. What were the subjects you would like to have read more about? _____

_____

4. What were the subjects or sections, if any, you would like to see omitted? Why? _____

_____

5. Were there any chapters or sections of the book you were not assigned to read? If so, which ones? _____

_____

6. Any additional comments or suggestions? _____

_____

Your name (*Optional:*) _____ Date _____

May Wadsworth quote you, whether in promotion for *Technology and Communication Behavior* or in future publishing ventures?

Yes _____ No _____

Again, my thanks,

FOLD HERE

CUT PAGE OUT

FOLD HERE

NO POSTAGE
NECESSARY
IF MAILED
IN THE
UNITED STATES

BUSINESS REPLY MAIL

First Class   Permit No. 34   Belmont, CA

POSTAGE WILL BE PAID BY ADDRESSEE

WADSWORTH PUBLISHING COMPANY
Ten Davis Drive
Belmont, California 94002